TIME AND A LIFETIME

© 2017 Vincent Calabrese
Published by Watchprint.com, La Croix/Lutry, Switzerland

Original edition *Le Temps d'une Vie* published by Editions Slatkine, Geneva
© 2016 Vincent Calabrese

All rights reserved. Any reproduction of this work in whole or in part is forbidden. Any copy or recording by any process including photocopying and photography and on any medium including microfilm, magnetic tape, discs or other storage devices constitutes an infringement of authors' rights punishable by copyright laws.

ISBN 978-2-940506-18-7

English Translation: Annabelle Larousse

Printing: Daneels Graphic Group, Belgium

VINCENT CALABRESE

TIME AND A LIFETIME

A Novel about one
of Switzerland's Most Innovative
and Creative Watchmakers

Table of Contents

I	Why I Wrote this Book	9
II	From the Stable to the Stars	13
III	Hello, Switzerland!	21
IV	So Now I'm a "Rital"	23
V	Further Education, Further Wandering	25
VI	Stepping Out on the Big Stage	29
VII	A Time of Disillusionment	39
VIII	The Real Turning Point	51
IX	Cloud Nine	59
X	The "Ludique" Years	65
XI	The New Millennium	71
XII	The Creative Spirit	77
XIII	Nouvelle Horlogerie Calabrese	79
XIV	Hard Times	83
XV	New Life	89
XVI	Corum	93
XVII	2015	97
XVIII	Rage	103
XIX	A Touch of Philosophy	107
XX	An Eye on the Future	109
XXI	My Two Lives	113
	Biography	115
	Glossary	121
	Acknowledgements	125

I
Why I Wrote this Book

Should I set out my life, my career, for all to see? Why not? For a long time, thanks to the media, a good bit of it has been known around the world. Still—and this is an important factor—apart from a few lectures I've given and some technical writings, it's always been told by others. Always in a complimentary fashion, to be sure. But always with the distortions that are part and parcel of the exercise, since they were talking about *my* life, not theirs!

Eventually those around me, my daughters in particular, asked me to write about it myself. A request that, as time went on, became more and more urgent since many people had very flatteringly done me the honour of taking an interest in my work as a creative watchmaker.

For a long time, I hesitated. It's a tricky thing talking about yourself. Always that balancing act between saying too much and too little. You run the risk of lapsing into boasting. I hope to avoid that mistake, to avoid giving the impression that my very diverse experiences in life have left me either too passionate or self-pitying.

This story will alternate between light and shadow. Though it might come across as modesty or boasting, I hope it will be an honest reflection of my life. I will not minimize or trivialize the facts in the hope of making them more palatable. But if I fail altogether in my task, I ask for the reader's indulgence and invite him to look upon my work impartially.

There are many reasons why I changed my mind. The least of them relate to two anniversaries coming up in 2017: the 60th anniversary of the start of my career and the 40th of my first invention, the "Golden Bridge". The other more fundamental reasons involve certain unexpected events. But it's not time to talk about that just yet! By the end of this book you'll know all about it. That will be the reward for your curiosity and your patience or perhaps your impatience to know how the story ends.

In this book, you'll often come across the word "philosophy". Don't worry, there aren't going to be any lectures on Aristotle or Plato or any of the other great philosophers who pursued *the* truth. I was more modest and only undertook the quest for *my* truth. I have never sought to convince others of anything. I have instead taken on the more arduous task of simply understanding.

If it's not too presumptuous, I hope that this book, with its account of my modest contribution to the history of horology and the crucial stages of my life, will be of service to others. I don't pretend it will change the world. The world will never change. People will always be people. The most I can do is shed some light on this magnificent profession and its evolution over the past few decades.

Perhaps this work will also be useful to the new generation that is looking to take up this profession. Perhaps the success and the hazards that I have encountered along the way will be instructive for them. Perhaps it will help them to remain honest and true to themselves, despite the pressure, the power struggles, the financial temptations and the narcissism that are as characteristic of our circle, the watch industry, as of all others.

I would also like to thank the reader for his patience with the technical discussions found in this book. Some of them are difficult to follow, even for those in the profession. So as not to overburden the reader, I have kept them to the strict minimum. Those who wish to know more about them can turn to the Internet, where there is plenty of information, not always correct, but comprehensive. One of the main features of my website is brief explanations of matters that might need clarification.

The Meaning of my Life

In short, I'm trying to find some meaning in what I've lived through in order to take what is important from it. Even as a teenager I clearly understood that I would not live forever. I hesitated to commit myself to life, until I decided that I preferred courage to cowardice. I was determined to make life interesting, to live fully while at the same time somewhat in the role of an onlooker, but always, at every moment, seeking to ensure the victory of the will, the spirit over the body.

Such determination has always kept me on the right road, even during the most trying times. I needed to keep my cool, not to give way to unreason. It allowed me to master my craft, to turn it into art and to make it my true language.

Not that I was aware of any of that at the start. It was only with the creation of my first watch that I capitalised on the freedom promised me by my victory over myself, over all the many urges that are normal at that age. A radical change was taking place within me. That first watch held a symbolic power that I didn't even suspect at the time: the mastery of time also involved that of my humble person.

What Is a Watchmaker?

We can say without a doubt that the current success of the watch industry is the stuff of legends. In the layman it creates a sense of wonder—though not always in keeping with reality. As in the past, this success continues to connect the image of the watch with that of the watchmaker as a craftsman. But this is deceptive. Why does this connection no longer reflect reality? What is different today? And what actually is a watchmaker at the present time?

In the 19th century, when industrialization involved the division of labour, master watchmakers were already sub-contracting part of their work. Yet they correctly reserved for themselves the sole right to put their brand name on their products. With the growing success of watchmaking, the practice of sub-contracting increased enormously, allowing anyone at all to put his own name on a watch. Unfortunately, the media, always needing a story to survive, were willing to grant such characters the necessary authenticity they were claiming. Finally, new technology, new raw materials and above all, information technology, provided them with everything they needed to award themselves the unearned title of watchmaker.

Now a watchmaker who is a true craftsman is much more than a salesman of his own products, whether he is surrounded by machinery or painstakingly devising dials and mechanisms. He is confronted by a world that is far more complex than that of the watch itself. His craft brings him into contact with a host of others. It demands of him an expertise, focus and strength of will that make the craft itself seem almost secondary: it becomes unavoidable—if he is endowed with the creative instinct and refuses to be a mere cog in the wheel—for him to turn to philosophy.

Creating a Watch

Every one of my creations has some significance. It tries to convey a message. Like a theatrical director, I attempt to harmonize the screenplay, the actors, and the dialogue in order to get that message across. Far from being a compulsory exercise, my creation is the response to instincts that rise from very deep within me. That's why I have no hesitation in speaking of philosophy. Besides the engineering and configuration of the various mechanisms, there is a purely existential dimension present. You must rise above the commonplace. Seek out your inner light. Knowing the time doesn't mean you know how to live the time that has been allotted you.

Creating a watch is a whole world. That of the watch industry in particular. Which can be the best or the worst of worlds. For once a watch is created, you have to promote it and sell it, a process involving a complex circuit that obeys the merciless laws of fashion and money. The world of watchmaking is hardly free from chicanery. You have to approach it with caution.

When it comes to my craft, everything involved in it and my relationship to it, I'm a self-taught man and glad of it. It was the best preparation for life I could have had. Not being guided by any mentor, not being a slave of the well-trodden path has made me a free-thinker with the right to question and challenge everything.

If a lack of formal education is a disadvantage, I have been able to profit from my know-how. The will to succeed is the best weapon you can have. It's also true that if I have had some success, it was due to this craft with its arcane nature which, finding expression in the narrow circle of the watchmaker's workshop, comes to fruition on the international stage.

II

From the Stable to the Stars

I was born in Naples on January 6th 1944. As was commonly done in those days, due to the wartime crisis, my birth was not registered until the 20th. I've always suspected that this seemingly insignificant fact influenced my whole life. Since January 6th was the Epiphany, I always got loads of presents then, which led me to believe that that feast day was actually in my honour. But this was no longer the practice by the time I was old enough to start school. This administrative error made me something of an exception at the time, and I could not share my secret with my little friends.

My brother, sister and I shared a 30 square metre room with our parents. My father was long unemployed. An ardent Communist, he was obsessed by a feeling of being exploited. As he was illiterate, violence was his only means of expression. My mother was 14 years younger than he. She was a gentle, sensitive and refined woman. Which is to say that the two of them were badly mismatched. Our living quarters were especially unpleasant, given that they were on the ground floor, right next to all the traffic and noise outside. In Naples, these apartments were called "the Bassi", since they had been turned over to those at the base of the pyramid.

Until the end of primary school, when I was almost ten, very shy though not unhappy, I took refuge in my own world. I occupied myself with my Lego set, constantly organizing battles in which, obviously, I always came out on top. One of my most poignant memories is that fateful day when, on my return from school, I could not find my Lego set. It wasn't where I always kept it. When I asked my father about it, he sized me up, his hands on his hips. "I threw it all out," he declared, totally unconcerned. "It's time for you to grow up. You're a man now!"

I hated my father, a violent man who never offered me any affection or praise. He was happy enough in that shabby apartment and with his circumstances, which he considered far superior to anything he had ever known before. When he was six, his own father had abandoned his family, and he had had to go to work in a bakery. In the end he became a highly respected master baker.

Just like me, he had hated his father and, because of him, the frivolousness of existence. He could not tolerate the least sign of weakness, amorality or dishonesty. These values, which he passed on to me, led him to reject the age-old tradition of naming his first son after his father. Instead, he named him after my mother's father. Five years later, when I was born, he bowed to family pressure and named me after his father. Another apparently trifling matter which nonetheless had a big influence on my life, though I did not immediately realize it.

My mother, the eldest of her family, lost her father when she was 14 and began looking after her brothers and sister. During the day she served as a companion to an elderly, upper middle-class lady, who gave her a taste for reading, classical music and good manners. In the evening she would come home to face the constant harassment of her stepfather, which became so unbearable that she finally left, taking refuge with my father, even though he was a friend of her stepfather's. She did not suspect that my father, who had sheltered and protected her so faithfully that she eventually came to love him, would prove to be as violent as the man she had run away from. True, my father was a violent man—but he was no less honest and generous. Above all, he was outraged by any injustice. He also took in my mother's sister and two younger brothers. It was undoubtedly because she felt indebted to him that my mother never considered divorcing him. But we have to remember the context of the times, when divorce was out of the question.

As time went on, relations between them became strained. My mother could not resign herself to such a narrow life. In January 1939 my brother was born, and in September of that year the Second World War broke out. She persuaded my father to buy a sewing machine so that she could make clothes for her children and mend those of her neighbours, thereby earning a little money to keep them going. This turned out to be a very good move. For in 1940 my sister was born, and when my father had to go off to the war in Africa, my mother was left to look after her children, as well as her sister and brothers, on her own.

A year later my father returned—only to leave immediately for Croatia, in the middle of winter, outfitted exactly the same way as he had been in Africa. He, like a good many in his battalion, suffered terribly in the serious cold, so that when he returned for good, unable to work

The street in Naples where I was born.

any longer as a baker, he had a long struggle before him to gain official recognition as a war invalid. I spent a good part of my early years writing letters of protest against the continued refusal to acknowledge his disability, which was finally officially recognized 15 years later.

The family survived thanks to my mother's labours. A self-educated woman, she became a seamstress, highly regarded for her made-to-measure clothing. For his part, my father reproached my mother for continuing to financially support her brothers, who were now grown. The tension between them was palpable. My mother and her siblings were very attentive to my brother, since he had been named after their father. They paid much less attention to my sister and even less to me, since I had been named after my father's father. On the other hand, my father made it perfectly clear that I was his favourite. Was that due to his natural feelings or to some unconscious spirit of opposition to my mother's family? It did not matter since I myself sided with my mother, to whom I devoted all my love.

She was ambitious for her children and worked away at her sewing machine from morning to night to provide a future for us. It was thanks to her efforts that my brother and sister were able to finish school. Gentle, refined, fond of reading and lyric art, she despised vulgarity, refused to let me mix with people of our background, and kept me away from bad company.

But I felt quite isolated. Since my sister was four years older than I, and my brother five, I had practically no contact with them. In general, I kept to myself and worked very hard to win the affection of my mother's family. I was determined to speak good Italian rather than the local dialect, to find friends that met my mother's standards and to prove myself worthy of them. This childhood experience forged in me a visceral need for recognition and a fierce will to succeed, since I was convinced early on that your start in life should not determine the finish. The fact that you were born in certain circumstances did not mean that you were condemned to them forever. It was possible to change them, if you had the will to do so.

My Early School Years

Once I had finished primary school at the age of nine and a half, I continued my compulsory education at a technical school on the other side of the city. Having been born in January, I was six months ahead and the youngest in my class. Some of my schoolmates were 14 or 15. Furthermore, I was very small and felt quite uncomfortable. I had almost an hour's walk to get to school. Since I didn't have the bus fare, I quite often managed to sneak a ride without paying. In the end, I had to repeat the year. My parents decided instead to send me to another school about ten minutes away. There it was a more classical education, which suited me much better, since the focus was on the arts, and particularly Latin, which I loved.

Bit by bit, I gained in confidence. Frustrated because I was not growing as quickly as my pals, I became heavily involved in sports, which made me a lot stronger, both physically and psychologically. I quickly built up an impressive set of muscles. But remembering how scrawny I had been before, I was always quick to stand up for the weaker boys and I knew how to get respect. I was small but tough!

The first year went extremely well, but in the second, tired of the slow pace of the classes, I began to play hooky. And so the inevitable came about: I failed. Another failure would get me expelled from school, and I would lose my right to an education. I had to change my ways. My parents would not have agreed to let me drop out, as that would have seriously compromised my prospects. I began cutting classes less often and worked harder to keep my grades up. Everything was going well. I was near the top of the class. Except that one day, everything fell apart. At the end of PE, we were going up the stairs to return to class. One boy was trying to push everybody out of the way to get in front. I told him to wait his turn. When he said I didn't know who he was, I replied that he needed to respect other people—whereupon I gave him a push in turn, and he fell down the stairs. It turned out he was the principal's son! Maybe it was a coincidence, but from that day on my grades, until then excellent, began to slip. And that's how I failed again and got expelled.

My Introduction to Watchmaking

It was now 1957. Forced to leave school, I was obliged to take a job as a "chemist's assistant" in a laboratory making pharmaceutical products. The boss was often absent, and I was on my own in the lab. Pouring a beaker of sulphuric acid into some other liquid one day almost cost me my hide. The boss was even more absent-minded than I was. He would forget to pay me. So I quit. But to show my mother I could get by, I landed a job as a delivery boy for a bookshop. Physically it was very demanding: I had to deliver parcels weighing up to 40 kilos—on foot, since transport was a luxury. It was well paid, but only temporary, around the time the new school year was beginning.

My mother rightly demanded that I learn a trade. A watchmaker's shop had placed an ad in the local paper. They were looking for an apprentice. My interview went well, and the boss agreed to hire me on condition that my mother bought my tools. So I began my apprenticeship and everything was fine—until a few weeks later when the former apprentice came back to reclaim the position. The boss was then at a loss with two apprentices on his hands. He decided to give the other lad his former spot on the bench and thereafter spent a lot of time teaching me various odd jobs.

It's true that during this apprenticeship I learned all sorts of things, like connecting a radio to an alarm clock or making metal chairs with a plastic cover. But none of it was terribly inspiring. Not to mention that the pay amounted to a pack of cigarettes a week. So I quit this "job" as well. But since my mother had bought me the tools I needed, I set up on my own repairing timepieces. I wasn't yet 14 years old. Repairing watches and clocks for my pals or for friends of the family earned me some pocket money.

My Introduction to Philosophy

As far as my personal development is concerned, there was one factor that was even more decisive than my formal schooling. As often happens at this age, my pals and I had turned our attention to spiritual (or perhaps esoteric) concerns. We eventually met a man who instructed us in various religions. He acquainted me with the value of being open, while giving us to understand that all religions were equally valid, that none was superior to any other and that, whatever else, they were an obstacle to freedom of thought. It was this budding awareness that led me to abandon Catholicism and become an agnostic. I no longer wanted to be the prisoner of any dogma. Thereafter, the focus of my concern was on the dominance of the spirit over the body.

I had got to know this spiritual mentor well, but I quickly forgot him when I began taking private music lessons with three of my pals. We decided to form a quartet. My instrument was the trumpet, which I loved because of its difficulty but hated because of the way it took over your life. A week without playing it and you would have to strengthen your lips all over again. While my friends went on to become professional musicians, for years afterwards I continued playing for my own pleasure. Music has always been one of my great loves.

Like most of us, I was shaped by my adolescence. The general nature of my character, my passions and my way of thinking became apparent at this time. Whereas I was immediately comfortable with certain aspects of that nature, it took me some time to recognize others, but there were none that I rejected outright. What I recall of my childhood is the boredom of the evenings spent with my family. Above all, there were no other children in my family who were my age. There was a difference of two years between me and the cousins nearest in age to me, whereas some of them were 40 years older than I. The result was that of all my years in Naples, the only time I felt I was truly living was from the age of 12 to 17, when I decided to emigrate.

Yet I benefitted a great deal from my independence. I was seldom at home. I would rarely come in before one in the morning, and sometimes it was much later. I was in open rebellion, constantly looking for a confrontation with my father. The money earned from my repairs or from seasonal work allowed me to get on nicely enough with my pals, who, though from wealthier homes, had less money than I since they were students. And since I'm recalling my adolescence, let it be said that I never saw anyone pull out a knife or a gun, nor did I ever witness any act of violence—only a few scuffles among young lads, during which no one was knocked to the ground. In Naples, the Camorra had disappeared at the same time as Mussolini. It would unfortunately make its reappearance in the 70s with the emergence of various corrupt leaders, the Red Brigade and the neo-fascists.

1961 Autumn
Departure for Switzerland

III

Hello, Switzerland!

The big turning point in my life was when I was coming up to my 18th birthday and was faced with doing my military service. Doing two years in the navy would be a nightmare, but the only way out of it was to emigrate. I thought about my uncle. He was living in Switzerland. Why not take the opportunity? In this respect, it's clear how much I was like my father: his ferocious need for independence, his refusal to submit—and his penchant for violence, which fortunately I had under control, and which nowadays has been diverted and channelled into my creative work.

Many Italians emigrated at this time. My mother's brother had left in 1960. There were other reasons I was thinking about him: although my own brother had finished his studies in aircraft construction and had found a position in Florence as a draftsman, and similarly, my sister had found work as a cashier in a bar and thus had achieved relative independence, the atmosphere at home was unbreathable. My father poisoned it with his idleness and aggressiveness. My mother was suffocating. So I got the idea that if the whole family took off for Switzerland, it would be the end of our problems, especially since my father, because of his war wounds, could not bear the cold!

I had no trouble in convincing my brother and sister. They left in June 1961. My mother and I joined them in November. As long as I live I'll never forget the day we arrived in Switzerland. I was leaving behind my hometown and my friends on a train packed with emigrants, knowing that I would perhaps never return. I shed a few tears: my mother, discovering that I had some feelings, was touched by them. At the border, the Swiss officials made us undress, and the in-depth medical examination was quite a shock. Another shock awaited us in Le Locle with its more than two feet of snow—and I was wearing only a light jacket. Compared to Naples with its population of around two million, this village of only 10,000 people seemed like a hole in the road. And yet I was happy. It was something new, something different, an adventure! Leaving my friends had not been easy, but I needed my freedom from a society that was too traditional and too oppressive. Le Locle appeared to be more tolerant, even though it was Protestant. In Naples your only choices were the Church and Communism.

Settling in Le Locle

At this time, finding work was no problem. A week after I arrived, I was hired by Tissot. My brother and sister also landed good jobs. We were easily able to afford a four-room apartment. I shared a bedroom with my sister. My brother, who was getting married, shared another with his fiancée, while my mother had her own. It was so much better than Naples where we were all cramped together in one room! So we had found a place to live and were quite content with our lot, especially since we were so far away from our father.

But at Christmas, everything turned upside down when he came to visit. After a few days—surprise!—he announced that he quite liked Le Locle and had no trouble putting up with the snow so he was planning on staying! Nobody was thrilled by that decision—particularly not my uncle, who made a point of accusing my father of sponging off my mother.

Then the inevitable came to pass. The following Christmas, in 1962, when we were all gathered together, my father and uncle came to blows. My old man was big and strongly-built. Lots of people were afraid of him. Fearing the outcome, I stepped in, doing what I had long dreamed of doing: I hit him squarely in the face and knocked him out. Somebody called the police. My father was arrested, tried and banned from the district. I instantly regretted what I had done.

Despite the cold, I went out on that Christmas night, crushed with remorse. I wandered around for a long time before I returned home, only then learning that the police had been called in. I began to see my father from a different angle. Thinking over what I had done, overwhelmed with guilt, I reflected on how he had once been a master baker, respected by all, then wounded in the war and made permanently unemployable. I saw his thirst for justice and honour, and his indignation at seeing anyone being exploited in a new light. I began to find some love for the man I had hated so much, feeling that I had been manipulated, turned against him by my family. I asked him for forgiveness and we were reconciled.

IV

So Now I'm a "Rital"

Making a start in Switzerland was not as simple as all that. At the workshop when I asserted that I was a watchmaker, the foreman sharply retorted, "No, you're an Italian!" And yet, if the times overall were not too good for the "Ritals", my background caused me little worry, at least at first. Like all my family, I was too happy to be in Switzerland. I quickly learned to speak French more or less correctly. A workmate lent me a copy of Zola's *The Kill*. I read it from cover to cover and could truly relate to it. In short, I felt that I was fitting in nicely, like all my family, except my mother who was having a bit of trouble learning French.

Still, I quickly became unhappy at Tissot. It was assembly line work, which did not suit my temperament at all. I sang while I was working, made jokes and took too much of an interest in my female colleagues. The foreman told me I needed to calm down. And furthermore, I should be wearing a tie. He didn't fire me, though, because my production rate was well above average.

I was tired of the atmosphere of the place, though, so I quit and took up work at another watchmaking firm, Cyma. There I enjoyed the benefit of a genuine apprenticeship. Although the business was clearly smaller with a more limited market, from a technical standpoint it was superior. The boss, a man by the name of Inäbnit, took me under his wing. In a few months we covered the complete process of assembling a watch. He was pleased with me. I achieved excellent results with even the most specialized aspects of the final setting, under the supervision of Racine, the head of the Correction Department. Thanks to his advice, I doubled the number of chronometer certificates we earned.

I was happy at Cyma. That was where I truly became a watchmaker and got an overall view of the craft. Yet I soon moved on, for several reasons. The first was my colleagues' jealousy. This was partly due to my competence, but also to my relationship with a colleague who commuted across the border (and who would become my wife). They thought an Italian had no business seeing a Swiss woman or one from just across the border and they left me in no doubt about it. Also, there was the prevailing hostility to foreigners. In the only place in Le Locle where a lad could go to dance and flirt with the girls, I had words with a Swiss man. The next day the story went round that I had pulled a knife on him. By the end of 1964, things had become impossible at Cyma.

My girlfriend and I applied for work at Zenith. As I have mentioned, at the time it was easy to get a job from one day to the next. Furthermore, there was no prejudice against moving around. On the contrary, staying too long in the same place meant you were stagnating. Determined to make a break, we became engaged while we were at it. The situation with my girlfriend's family was not exactly rosy, either. Her mother did not like me, and her uncle could not stand the thought of her dating a "Piaf". The lone exception was her father, himself a watchmaker, whom I got on well with and for whom I had a great deal of respect. My love affair with Le Locle was seriously on the wane. As hard as I had tried, I had not succeeded in making a single Swiss friend. Although I could understand the fears of this little town in the face of the invader, I much preferred La Chaux-de-Fonds where relations between the locals and foreigners were more relaxed.

V

Further Education, Further Wandering

My job at Zenith did not suit me. I had been promised that I would be assigned to the Repairs Department rather than carrying on with assembly-line work. Yet ten months later I was still in the Setting Department because I was quite competent there. In December 1965, I got married. On our return from our honeymoon, desperate to get out of Le Locle, I found happier circumstances in the Richard factory in Morges. This position allowed me to do repairs. Moreover, the company provided us with an apartment. Richard, which owned a chain of stores, sold and repaired all brands of watches, which allowed me to gain a working knowledge of the most diverse aspects of the craft.

Family Life and Marital Problems

Unfortunately, problems in my marriage arose at precisely this time. Bit by bit, my wife and I were drifting apart. All of a sudden, one day she decided she would stop working and I had to shoulder the whole responsibility myself. I buried myself in my work. In a way, I have to be grateful to her. Without these marital difficulties I never would have found my path in life. My thinking also began to evolve as I became active in the trade union movement. I organized an Italian club and was named its president. I also began giving Italian lessons to the Swiss and French lessons to the Italians at the Migros School.

The birth of my daughter Marina in January 1969 was one of the most beautiful moments of my life. She gave me one more reason, if I needed one, to keep trying to move up in the world, to make a better living so that I could offer her a better future. Yet (nothing new) I became disenchanted with Richard. Despite the fact that I was the newest member of their staff, I was amongst those who were the most productive, particularly as regards the most complex mechanisms. The management was satisfied with my work. I had even solved a fine little technical problem, which was going to lead to a pretty fair reduction in our production costs. The manager thanked me in front of the staff and announced that in future, the procedure I had devised would be adopted. Since I was at the bottom of the pile in terms of salary, I took the opportunity to ask for a raise.

Management did not follow up on this however. They even refused to implement my new procedure. Finally, some drastic changes in our work conditions forced me, along with five of my colleagues, to hand in our resignations. It came at the wrong time. It was at the end of 1969, Marina wasn't a year old, and there was no other work for a watchmaker in the area. I had to move and return to the canton of Neuchâtel. But I was sure I would find something else.

Departure from Morges and Further Technical Education

I quickly got a new job with Hebdomas in La Chaux-de-Fonds in the after-sales department. This was not a terribly interesting position. However, it was a living and it allowed me at the same time to enrol in a night course for operations technicians at the local technical school. This course was intended for those who had earned their CFC, a diploma in watchmaking, mechanics or electrical engineering. My ten years in the profession entitled me to enrol in the course. During the day I devoted my energies to improving productivity at Hebdomas. At night I learned trigonometry, physics, algebra and electrical engineering, subjects that I had never studied before. I loved it, fully aware that what I was learning would be essential for my further advancement. Although the first few weeks were heavy going, given that I had left school when I was 13, I caught up, so that at the end of the first year I found myself near the top of a class of some 30 students. What makes me even prouder is the fact that at the time there were no calculators or computers. All our calculations had to be done with a slide rule!

First Challenge, First Act of Rebellion

After a year, realizing that I had no real prospects at Hebdomas, I applied to Teriam, who were looking for a shop foreman. This firm was devoted to the creation of exclusive, private label models, but in greater numbers than was the case at Hebdomas. It employed around 70 people and specialized in budget-price watches for the Near East. I was put in charge of the department concerned with the fitting of dials and hands.

The Technical Director was satisfied with my results. I also agreed to head the Casing Department, concerned with fitting the completed movement in the watch case. This being the final stage in the assembly of a watch, it required care, method and preciseness.

This department was not the most efficient in the firm and the technical director asked me to improve on productivity and quality. It was the type of challenge I liked. He promised to assign me an assistant and to give my staff a raise if I succeeded. Well, I got no assistant, although every other department had one. But that did not prevent me and my staff from doubling productivity. Chuffed by this success, I demanded a raise for my workers, which was not approved. The boss argued that the firm had always got by at the same wage levels before I came along, and they would get by after I was gone. Since he was refusing to budge on the issue, I pointed out that I was trying to keep my promise to my staff, but he was not. And I handed in my resignation.

This was in December 1971. In November, my wife had given birth to my second daughter, Tina. In the circumstances, I have to admit that my resignation was a bit cavalier.

View over Crans-sur-Sierre

VI

Stepping Out on the Big Stage

A few days later, during the Christmas holidays, I saw an ad looking for a watch repairman capable of managing a shop in Crans-sur-Sierre. I was happy in La Chaux-de-Fonds and had made lots of friends. I could look forward to a peaceful life there, with no more need to move around. And yet the prospect of working at my craft as a repairman with almost double the salary, in a fashionable resort in Valais, a canton with a more pleasant climate than the Jura, was more than tempting.

Aeschlimann's jewelry business handled the most prestigious brands. This watchmaker who hailed from the Jura owned one shop in Sierre, one in Anzère, two in Crans and another in Montana. "My shop", one of those in Crans-Montana, was appropriately named "Le Diamant Bleu". It was the most profitable of the group. I moved to the resort in February. It was a huge change. Besides the increase in pay, I had six weeks' paid vacation along with my health insurance.

So there I was, a watchmaker-manager! The best of all possible worlds. Most of our clients were from Belgium, Paris and Milan. My Italian came in handy. I was the only one in the shop who could speak it. I will always remember my first day on the job. I was in the middle of organizing my tools when the boss called on me to look after an Italian customer. I showed him a few watches and succeeded in selling him a Patek Philippe costing 8,000 francs—quite a price-tag at the time. (A similar watch would cost around 25,000 francs today.) The boss very quickly handed over a lot of responsibility to me: opening and closing the shop, seeing to purchases and acting as manager in his absence.

I got along very well with Aeschlimann. Furthermore, between my salary, my share of the profits and incentives on unsold items (nobody thought much about the unsold watches, but I did! I was a good salesman and thought it better to miss out on a sale rather than lose a customer), I was earning between 5,000 and 6,000 francs a month.

These were wonderful years. Besides my enjoyment of the work and my good relations with the boss, there was the additional attraction of the area. Crans-Montana was frequented by high society. I often served celebrities—actors, singers, politicians, even members of royal families. I was very relaxed in my dealings with all my clients, my Neapolitan side always putting business before my personal concerns. Some of these personalities made a point of dropping into the shop to say a quick hello or have a friendly chat even when they did not need anything in particular.

It was all very flattering. There I was, a little Neapolitan from the humblest of backgrounds, and finding myself in such exalted company made me think I was part of it. It never occurred to me that these people were only interested in my services, not in me personally. It was only much later that I became aware of how deluded I was—just like in my adolescence when I mixed with young lads from more comfortable circumstances that I dreamed of enjoying someday.

The Trigger

This was nonetheless a pivotal period in my life. It was then that my creative instincts helped me find my path. My desire to create was not new. It went back to my time at Hebdomas, when one of my colleagues was so proud of a patent he had been awarded. I envied him so much! Nothing fascinated me, a self-taught man, so much as the thought of a patent on my own invention. Hence, the impulse to create. The desire to make a lasting impression. As a teenager, I was aware that I could do without music and sports, just as later on I could do without the social success I thought I enjoyed in Crans. I was feeling unfulfilled. When he was 20, didn't Caesar weep because he had not achieved what Alexander had when he was 18?

At the Diamant Bleu, our clients were well-off, concerned about their image and always wanting a personalized watch. Although our shop offered every possible brand, and the most prestigious, too, they coveted something unique. I turned to the manufacturers, who informed me that at best they could engrave the case back or design an original dial. In a word, they could produce nothing personalized, though that could be done even in China or Naples. My clients were disappointed.

Then one day a regular customer called, accompanied by his wife. She showed me a splendid Breguet pendant watch from the 19th century that had been run over by a car. They asked me for an estimate on the repairs. I had a look at it. The movement could be repaired. The case, however, was seriously damaged. I passed on their request for an estimate to the jeweller in one of our shops. The repair of the movement—the watchmaker's concern—would cost about 800 francs (around 1,500 francs in today's terms). The repairs on the watchcase—the jeweller's concern—would come to around 2,000 francs (4,000 francs today). My client asked me just to repair the case!

What? He wasn't the least bit bothered that the watch wouldn't actually work. He was only concerned about its looks. I was shocked. He pretended to love watches, but couldn't care less that the movement was engineered by one of the greatest watchmakers in history! That told me that my craft counted for nothing. It meant that a jeweller could earn 2,000 francs while a watchmaker could not even earn 800. This realization produced a major tremor inside me. From that day on, I was determined to create a watch that people would buy solely for the beauty of its movement as soon as possible.

This episode filled me with a desire for revenge. The watchmaker's craft is demanding and thankless. Especially when it comes to repairs. You have to take a watch apart, and understand its workings before you can even begin, whereas the jeweller has nothing more to do than to solder or file down various parts. Even at Richard, I had run into this injustice. The jewellers there earned 20 to 30% more than the watchmakers. Our craft was undervalued.

Towards a New Concept of Watchmaking

I was filled with rage. Of course, I had it all, I was earning a good living and I had two wonderful daughters. But I was gripped by a sense of uselessness, of emptiness, of meaninglessness. All this had come over me so suddenly, as if my craft and the watchmaking industry were only a matter of smoke and mirrors, without any solid foundation. I wanted to go further, test myself, and prove to myself what I was capable of. It was the exact opposite of what my life had been up to then—the pursuit of recognition, of a standing which, in the end, had not got me anywhere.

I now wanted to try to create something that came from me alone. I was haunted by feelings like this when I decided to buy a watchmaker's lathe. To be sure, it was bottom of the range, the type you can pick up in a department store. It cost me 500 francs.

In the mid-70s the oil crisis and the fall of the French franc had caused a serious decline in business. Furthermore, the introduction of electronic watches was a critical challenge to mechanical watches, and the number of people employed in the industry fell from 120,000 to less than 20,000. My profession, if continued in the traditional way, was going to die. It was not my nature to sit back and see what would happen. So I decided to specialize in the restoration of old watches.

My boss knew all about it and he had no problem with it. My salary, comfortable but barely sufficient for a decent living for my family and me, would not allow me to buy any machinery. Taking courses in watchmaking was also out of the question, since there was no school for watchmakers in Valais. With my little second-hand lathe I was at first content to get an 18th century fusee watch back in working order. I managed to replace the damaged verge with another one. This was a question of fitting, rather than of manufacturing, since I had only had to shorten it somewhat. Still, I was pleased and proud of this initial success.

However, I realized that I was not suited to restoring watches. Every time I took on a new job, I wanted to improve the item after my own fashion. Fortunately, I was principled enough not to do so. But I was either going to have to find a different sort of work or carry out my own in a different way. I then decided to start creating my own watches. Until then I had done no more than submit ideas to various people whose sole response was: "Unrealistic". I was miserable. To be sure, I was excellent at repairs, more than competent, but I had no theoretical knowledge of the manufacturing process. Moreover, though I had some mechanisms of my own invention in mind, my lathe was too limited for me to actually make them.

Further Training at Patek and Rolex

I used my six weeks' paid vacation to take the first step towards solving my problem. I divided the time in two—three weeks spent with my family, then three weeks devoted to training courses with the manufacturers that supplied our shop. If several of these courses taught me nothing, two of them were immensely rewarding.

The first was in 1975 at Patek Philippe. They ran their repairs workshop in the traditional manner. Their craftsmen possessed skills that you could only dream of, and their spirit of cooperation was exemplary. During this three-week course, I learned how to make various parts, such as a pull-out piece, a winding-stem and a balance-staff completed by a pivot-maker.

It would be difficult for me to explain these matters to the uninitiated, but a watchmaker or a knowledgeable connoisseur will certainly be impressed. It was obvious that my recent experience in restoring old watches helped me gain a lot from this course. The only snag was in the

Setting Department, where the manager tried to prove that his way, the traditional way, was the only valid one and that I had strayed into heresy.

The second training course, at Rolex in 1976, was a completely different experience. The shop foreman annoyed me for the entire three weeks, going on and on about how in their shop, everything was done right, and a part should never be repaired, only replaced. He was forgetting that I had been one of their dealers for the last four years. I knew their work practices perfectly well. By the end of this course I was appalled by such obtuseness.

At Patek Phillippe, on the contrary, I learned something about modesty. Their staff went about their work very knowledgeably but with restraint. I was happy to meet colleagues who knew so much more than I did. Nonetheless, I was still unhappy to see that despite their mastery of their craft, they stuck to the old rules so solidly established that I believed they hindered the true spirit of creativity.

I was convinced that innovation required the courage to leave the well-beaten path, for it's a fact that there's an exception to every rule. I would have to find my own way.

Schwarzenbach's Collateral Effects

Meanwhile, tension between immigrants and the native Swiss was at its height. True, the "Schwarzenbach Initiative", aimed at reducing the numbers of foreigners, was rejected by 54% of the vote. Still, it was a profound shock to foreigners long settled in Switzerland, and many of them decided to return to their own countries.

My sister and her husband went home, but my mother decided to remain in Switzerland and came to live with me in Valais. I gave her the room where I had previously had my workshop, which I then set up in the bedroom my wife and I shared. I can still hear my wife joking when she found filings in the bed!

The First Calabrese Creation: The Golden Bridge

My recent experience had strengthened my determination: I had to create my own watch! I had learned my craft inside and out, but I had no diploma to show for it. I decided to set a test for myself in creative watchmaking, freely and purely for myself, so that I could feel fully entitled to call myself a watchmaker.

Clearly, with my primitive tools, I could not aspire to create a masterpiece. I thought I could start with simple inventions, perhaps a small clock, even a movement for an alarm clock or pocket-watch. To start with, I needed some metal, though I neither knew what sort nor where to buy it. I began looking about for the raw materials I needed. At Aeschlimann's shop we received a case of watches which was fastened by some large copper staples. I thought about using them. I took them all, while noting that they were too narrow to be of any use for a watch movement, even for a ladies' watch. Furthermore, the copper was so soft that you could bend it with your fingers.

So I came up with the craziest challenge ever: to create a movement that no one had ever seen before, to create the personalized watches that my clients were always seeking but which had never been produced. To be sure, there were famous movements dating from the 16th and 17th centuries, in the shape of a mandolin, a cross or a skull. But these movements were solid constructions with the gear-train concealed. My movements would consist of initials, symbols or any other figure the client requested. Above all, I wanted the entire gear-train and escapement to be visible so that people could admire the beauty of the mechanism.

I decided to call this new genre of watchmaking "Spatial". I wanted to make the movement independent of the watchcase so that it would be free in its own space, which would allow me to place any figure at all therein. In other words, I wanted to cut the umbilical cord between the movement and the rewinding and time-setting mechanisms. It should be clear that the challenge no longer lay just in my ability to make a complete watch from scratch.

However, as the Italian proverb has it, there is an ocean between saying and doing! What was needed was something truly innovative, a movement that had never before been conceived or dreamed of. Early movements had been rewound with one key, while the time was set with another. At the end of the 20th century, I could not imagine returning to that state of affairs. I owed it to myself to devise a system that was simple and user-friendly. Night and day for nearly two years, there was only one idea in my mind: I was obsessed with finding the solution to this problem.

After almost two years, when I woke up one morning, I had it. A daring solution, even for a well-established manufacturer. As if in a trance, I began to devise this system that would permit the creation of my spatial watches. I held the solution in my mind, like an image materialising before my eyes. In my bedroom workshop, I at last began to design it, to drill through the metal, construct the gears and to fashion the watchcase. And all that with tools two centuries old.

Not only had I passed the test I had set for myself, but I found myself travelling a road that could properly be called philosophical. I was filled with the conviction that I could at last begin to express my truth concerning horological matters. And, who knows, perhaps concerning life itself. When my watch began ticking away, I began to cry. As when my daughter was born, silently the tears began to flow and I could not stop them. I realized then that my life was going to change.

I realized that my concept of watchmaking, with its freedom in space, offered limitless variations. Any figure can be drawn with a line, which can be straight or curved. So, recalling the thin staples I had taken from the case, I fashioned the world's thinnest watch, as narrow as a match, endowed with a movement suspended in space, without a dial which conceals the watchmaker's skill. Those around me were amazed. My wife, of course, but her family as well, including her uncles, who were also in the business. My brother-in-law was a prototyping watchmaker at Jaeger-LeCoultre. When I showed him my watch, he looked it over from all angles, and then whistled with admiration, declaring, "This is a bombshell!" That's when I knew I was on the right road, for he was very sparing with praise.

I needed to apply for a patent to protect my invention. I was familiar with the procedure since I had never been able to pay an engineering consultant. My brother helped with the technical design. Later I would also take on this task myself. I could already see myself inundated with orders from jewellers who would make expensive watchcases to house my unique movements.

It took off from there. I showed my baby to the curator of the *Musée international d'horlogerie* in La Chaux-de-Fonds, and he was full of praise as well. When he asked me what I planned to do with my watch, I told him my hope was to find a manufacturer who would buy my prototype. Once I had my patent, I knew I would be free to create my own watches. "I know someone," the curator told me.

In fact, he was friends with René Bannwart, the owner of the Corum firm. Three hours later, the contract was signed! I had sold the manufacturing and production rights, on condition that the movement would be made of 18-carat gold. There would be no dial, and the case would be completely transparent so that the movement could be viewed from all angles. As the movement was mounted on a single bridge, Bannwart suggested calling it a "Golden Bridge".

In his enthusiasm, he glimpsed the possibility of fashioning the entire watchcase in transparent white sapphire. Such was the birth of my "Golden Bridge"! Though at the time I considered it a lesser child amongst my creations, 40 years later it remains an icon, admired throughout the world. I had achieved the impossible: I had created a watch that people would buy for its beauty and the apparent simplicity of its movement. The Golden Bridge has become a flagship of the industry. The Maison Corum can take pride in offering one of the best-selling watches in the world.

Winning Recognition

A few months later, in November 1977, the International Exhibition of Inventions of Geneva was held. My prototype won the gold medal. At last my name would be known, money would come in, I was taking flight. At this time, all the great watchmaking firms were in hibernation. René Bannwart realized that with my unique movement his company had every chance of becoming a big player in the industry, which later proved to be the case.

Sure that I held my fate in my own hands, I left Aeschlimann. I decided to leave the mountains with a view to setting up in my own right. So began my initiatory journey.

1977 Prototype of a Spatial Watch

Specifications: movement in nickel silver. Length 29 mm, width 2.3 mm, thickness 3.2 mm. Setting of the time and rewinding of the spring by means of the barrel arbor. The peculiarity of no apparent link with the rewinding and time-setting mechanisms allowed for the crafting of movements of any shape, since they were independent within the space provided by the watchcase.

Noting that watches were in fact appreciated more for their case than their movement, I decided to reverse the situation and to create a watch admired primarily for the beauty of its movement and secondarily for its case.

1980 Golden Bridge

The first of my Spatial watches produced by the Maison Corum, the Golden Bridge, was officially christened in 1980 at the *Musée international d'horlogerie* in La Chaux-de-Fonds. Its instant success, due to the beauty of its movement, continues today, 40 years after its creation. This watch has become an icon in the world of watchmaking since the beauty of its movement is not concealed beneath a dial but remains visible through a transparent sapphire case. The utilization of a traditional case would have cast a pall over the watchmaker's ingenuity.

Golden Bridge Front and Back
Watchcase in 18ct yellow gold and synthetic white sapphire.

VII

A Time of Disillusionment

I decided to settle in Lausanne. I preferred it to Geneva, which I knew well from the training courses I had done at Rolex and Patek and which had less character. In Montchoisi, a working-class district of Lausanne, I discovered a jewellery shop that had closed down. What with my savings and the 20,000 francs I had got from the transfer of my patent to Corum, I was able to put down the rent deposit. I was happy to rediscover my roots after a fashion in this modest neighbourhood.

Early on I realized that this would not be a rose garden. Actually, to be honest, it was a nightmare from the very start! My turnover barely covered the 600 francs rent. In the first month I only sold a few watches and rare jewels. Furthermore, I was feeling extremely isolated since my daughters, then eight and six years old, had stayed in Crans so as not to have to change schools. I only saw them on weekends.

For three years, from 1977 to 1980, I earned next to nothing. After a year, when I still was not earning a living from the repairs I did for various shops, I decided, like the self-taught man I was, to learn jewellery-making. The idea came to me quite naturally since the place had formerly been a jewellery shop. Well, I have to say that these hard times had a positive, in fact decisive, influence on later events. Especially because, forced to learn this new craft to get by, I learned some valuable things about the analysis and utilization of gold, the handling of fire, gemsetting, enamelling and engraving.

Since, for lack of money, I could not subcontract any of the work, I had to do everything myself. Eventually, I realized, in my isolated, little retreat, that my prestigious clients in Crans-Montana had never been interested in my humble self or my talents; they only dropped in to see the manager. They had forgotten me and never showed up in Lausanne. A fine lesson in humility. What I call "business-card syndrome", the advertisement of your services along with your self-image and your illusions.

It was at this time that I began calling everything into question. Up to then, the road I was embarked on was closer to seeming than being.

This stark realization allowed me to recentre myself and to understand that my independence, whatever it might cost, was nonetheless my greatest asset. I took this revelation as a gift. A disaster can finish you—or it can give you the start you need. Still, I had a tiger by the tail. Fortunately, my wife finally decided to go back to work. It was in 1977 that I sold my patent, but it was not until October 1980, that Corum finally brought out the Golden Bridge and began paying me my long-awaited royalties.

First Difficulties with Corum

Corum finally sent me a few orders for Spatial watches. Things seemed to be falling into place. Except that in 1981, although our contract provided for the mention of my name in any publicity connected with the Golden Bridge, Corum wanted to revoke this clause, pleading that my Italian name was too hard to pronounce! By way of compensation, the company offered to increase my royalties. But this did not stop them, several months later and without asking my opinion, from using my name in their publicity for a magazine aimed at collectors, amongst whom my name was more esteemed than Corum's! Irrespective of this, I carried on doing repairs and creating my jewellery.

The International Competition in Le Locle

In 1982, the *Musée d'horlogerie du Locle* announced a competition to celebrate the 25th anniversary of its founding coming up in 1984. The requirement was to create a pendant watch with analog display, and you were allowed to submit three proposals. Feeling I was up to the challenge, I obtained all the details and submitted my three proposals. The first was in an elliptical case, with a linear movement and a tourbillon escapement. At the time this sort of design was extremely rare. In the two centuries following the invention of the complication only about 600 such models had been produced. (Today they're made at a rate of 600 a minute!) The second, in a shield-shaped case, was one of the Personnelle range with a movement in the shape of the letter M. The third, in the same sort of case, was a Personnelle in the shape of an H.

For those wondering what a tourbillon is, it is a modification of the escapement designed to improve the precision of mechanical watches. To explain further, its purpose is to compensate for perturbations in the balance wheel's isochronism due to the effects of gravity. It was invented in 1801 by the French watchmaker Abraham-Louis Breguet.

In 1983 my proposals were returned to me. Disappointed, I inquired what sort of innovation had been chosen ahead of my tourbillon. "Dear Sir," they wrote back, "we have judged your project to be impracticable. We know of no one capable of producing a tourbillon, particularly of the size specified." In May 1984, it was announced that the first prize had been awarded to a jeweller from Vaud. He had created a very futuristic pendant watch, endowed with a magnificent and visible movement. This movement, which had been supplied by Corum, was none other than a Golden Bridge. So I flattered myself that in this indirect fashion I had won first prize!

1982 Personnelles

The first examples of my Spatial watches with movements in the shape of letters were wristwatches or pendant watches, in traditional cases or of transparent sapphire, to display the beauty of the movement to its best advantage.

Personnelle S and Personnelle M

Specifications: movement in 18ct gold, size according to shape, Thickness 3.2 mm. Time-setting and spring-winding by means of the barrel arbor. Watchcase in 18ct yellow gold and synthetic white sapphire.

Founding of the *Académie Horlogère des Créateurs Indépendants*

Incensed at the disdain shown to my profession, at this time I remembered my colleague Svend Andersen. I had known him for many years. He had founded an organization for Genevan craftsmen watchmakers, though it had later disbanded because of internal dissension. But then I had the idea of founding an international organization.

In 1983 I contacted Svend and informed him of my project. The purpose of the organization would be to increase the visibility of independent creators and to seek sponsors. It would be an international organization, open to all craftsmen watchmakers, and a counterweight to the big manufacturers. On the strength of the success of my Golden Bridge, I asked various parties in the specialized media to carry an advertisement in the hope that other craftsmen watchmakers would join me. It appeared in several countries. Svend's support was invaluable since he spoke several languages, and we were able to get organized.

In 1984, a dozen or so craftsmen met in Lausanne to lay the foundations of our association. I suggested that we call it the *Académie Horlogère des Créateurs Indépendants* (AHCI).

There were two other craftsmen watchmakers that I hoped would join us, the only ones, to my knowledge, capable of producing their own watches—the Englishman George Daniels, known throughout the world, and the Frenchman Dominique Loiseau. I only contacted Loiseau, who agreed to join and announced his approval of the AHCI's first exhibition in 1985. However, he dropped out at the last moment. It turned out he had just been hired by Omega to produce the clock called "La Rose des Temps". This was treason on two counts. Not only had he abandoned me two months before the exhibition, he also officially announced, in co-operation with Omega, the launch of the International Grand Prix for Young Watchmaker-Creators. On the other hand—balm for the soul—Daniels joined us in 1988. Despite these difficulties, the AHCI saw the light of day and met with great success. Today, 31 years later, it's still going strong.

Exhibition of my First Tourbillon

On June 29th 1985, at a preview of the AHCI exhibition, by way of thanks for the confidence shown in us, I eagerly exhibited my "impracticable" tourbillon. When I created it, I could not envision a balance wheel above the escapement, with the hole in the center of a revolving cage.

It was too complicated visually. I was never a fan of the complication, whether it was functional or purely aesthetic. And I rarely saw anything in a skeleton watch or a tourbillon cage. I found such overlaying of lines and moving parts distasteful.

What was required was a more complex balancing of the cage, which would allow a linear arrangement of the balance wheel and escapement so that the functioning of the mechanism could be observed in its entirety. One advantage of the self-educated man is that he's not afraid to break taboos. My first tourbillon had been equipped with a steel cage, which was long-established practice—and absurd. For two centuries watchmakers recommended that the tourbillon cage should be as light as possible. They therefore used tempered steel, which allowed for a high degree of rigidity even with very thin supports. But the mere presence of steel, a highly magnetic material, was contrary to the whole point of the exercise: to increase the watch's precision.

To get around this absurdity, I fashioned the cage out of another, non-magnetic metal. For my first efforts I used 18-carat gold, and I decided to make my cage out of the same metal, though it was twice as heavy as steel, in order to show that the weight of the cage was unimportant, provided that it was well balanced.

1985 Tourbillon

First Production of a Tourbillon on the Occasion of the Founding of the AHCI.

This rather unorthodox conception encouraged me to maintain my independence and proved the value of calling into question even well-established principles. However, if the creation of my Spatial watches had aroused some jealousy, the tourbillon won me much outright hostility. A self-educated man, I was now becoming an iconoclast. Some thought I was completely misled: my tourbillon was in reality only a karussel, in other words, nothing much. I spent years proving that they were totally wrong.

Initial Collaboration with Blancpain

Nonetheless my work was making an impression. I had already implemented this second innovation when in late 1985 I was contacted by Jacques Piguet, director of the House of Piguet as well as of Blancpain,

Specifications: movement in 18ct gold, size according to shape, thickness 4.2 mm. Time-setting and spring-winding by means of the barrel arbor. Watchcase in 18ct yellow gold or synthetic white sapphire.

to discuss the possibility of collaborating, specifically on the development of a tourbillon for Blancpain.

I suggested two options to Piguet. The first was a classic tourbillon, in the tradition of Abraham-Louis Breguet's invention—that is to say, with a cage that rotated once per minute and a balance wheel positioned at the center. Or secondly, a tourbillon of my own devising, with an off-center escapement—though I was aware that the purists would denounce it as a karussel, a complication that was generally despised. He opted for the latter. Two months later, I delivered a prototype with a flying tourbillon—not just a technical achievement, but also an aesthetic one.

In order to produce the most impressive tourbillon possible, my design called for it to be on view not just through the dial, but through a crystal case back to allow the movement to be seen. With this in mind, I needed to fashion a cone-shaped aperture in the main plate, while the inner walls would be given a mirror-polished finish to allow the tourbillon the maximum amount of light. I had also insisted that it be placed at the top part of the dial, at 12 o'clock, the spot generally reserved for the manufacturer's logo. All of this was done for the purpose of ensuring that the mechanism would be visible when the watch was displayed in a showcase. Otherwise, if the tourbillon had been located at the bottom, at 6 o'clock, the oscillating weight would have obstructed a clear view of it.

Blancpain Prototype Front and Back

Obverse

This design also allowed for the possibility of a long power reserve. Thus the "eight-day watch" was born, also the thinnest ever made. The possibilities of additional mechanisms on both sides of the watch had been allowed for and were implemented as time went on.

This tourbillon was 3.2 mm thick and ran for eight days, a record that still has not been equalled. I sold my working prototype for a very modest sum in exchange for the royalties to be paid on each sale for as long as the movement was being produced. In the entire history of watchmaking, there were only about 20 wristwatches with tourbillons, and none had been produced for over 20 years. My wristwatch with a flying tourbillon was a first. I advised Jacques Piguet to take out a patent on the mechanism. His reply was that it wasn't necessary. Nobody would be able to imitate it, he said. And then in 1989 the first tourbillons from Breguet and Roth came on the market! And in the late 90s the first Chinese tourbillon did me the honour of copying my unpatented cage.

Still, once again I thought that we were getting somewhere—until the House of Piguet, which had undertaken to produce this watch with a tourbillon, let me know that my invention had been relegated to the background. And why was that? In the meantime, Jean-Claude Biver, an associate of Jacques Piguet, had been in touch with Dominique Loiseau concerning the development of a highly complex watch. He had succeeded in convincing Piguet to give this watch priority, since in his view it had more potential commercially. The intention was to produce 30 of these watches named "1735", the date of the establishment of Blancpain, and to offer them to subscribers at a cost of one million francs each. All of them were sold. But this project caused Blancpain a world of problems, since, as it turned out, the watches were highly unreliable. The company had to rethink them from scratch. The last watch of this series was delivered after a huge delay, long after its purchase.

Two years later, at the Basel Fair, I proceeded to the Blancpain stand, where Jean-Claude Biver was doing the honours. I asked him for the latest on my tourbillon. "What tourbillon?" he replied with feigned surprise. "This one?" And he showed me the "1735" on his wrist. I let him know I wasn't the least bit happy and walked off.

So it was a repeat of the difficulties I'd had with Corum. I had been counting on the launch of the Blancpain Tourbillon, but had to wait until 1991 for production of my watch to commence so that I could start collecting my royalties. But, as Freud said, isn't delayed gratification more gratifying? To gain real recognition, I was obliged to wait until 2001 when Marc Hayek was named to replace Biver as CEO. He was eager to be filled in on the history of this tourbillon and to meet with me and thank me for the creation of this watch.

First International Recognition

What doesn't kill you makes you stronger. It's true. My intransigent nature is reflected in my work. When I take on some task, I give it my all to the very end: that's my main strength. I have never started something that I didn't finish. I stop breathing, and time and space cease to matter.

A TIME OF DISILLUSIONMENT

Wherever I might be, day or night, I pursue my goals, I anticipate what will come. Time has no more hold on me, I dismiss it entirely. I work quickly, perfectly well organized. Before I even sit down on my bench or slip on my magnifier, before the machinery even begins to turn, I know what I need to do. Whereas in day-to-day life, I can sometimes be absent-minded or clumsy.

In 1986, while waiting for Blancpain to start producing my tourbillon and while my Personnelles were selling poorly, I set about creating a base model for my Spatial collection. I needed a motif that would please everyone everywhere in order to encourage sales in different countries. With this in mind, I decided to utilize figurative art, without however venturing into abstract art, for the creation of an allegorical watch that would express the spirit of my Spatial watches.

I envisioned a movement consisting of two arms starting from two opposite points. The one starting at 6 o'clock would comprise the barrel and its spring, hence its energy, thereby representing creative energy. The opposite arm, starting at 12 o'clock, would comprise the escapement, representing the mastery of technology.

It is always hard to marry art and technology, but in my "Esprit" concept, this is achieved for a brief moment by means of the hands, to signify the brief moment we live before departing into the infinite.

1986 Esprit

Specifications: movement in 18ct gold, size according to shape, thickness 3.2 mm. Time-setting and spring-winding by means of the barrel arbor. Watchcase in 18ct yellow gold and synthetic white sapphire.

Tourbillon Maçonnique
Watchcase in 18ct yellow gold and synthetic white sapphire.

Also at this time, two countries did me the honour of recognizing my work. In 1986, the Association of Belgian Horologists invited me to give a lecture in Brussels. That earned me a commission for a personalized tourbillon, the "Maçonnique". Then the Association of Spanish Watchmakers flattered me by awarding me first prize in a competition they had organized. The prize was a gold clock, which I unfortunately had to melt down in order to pay for my plane ticket. But the memory of the event still remains!

First Time for the AHCI at the Basel Fair

With my fame spreading, thanks to the success of the Academy in 1985 in particular, I contacted the Basel Fair to suggest that they invite the AHCI. My request met with a favourable reception. The Academy made its first appearance in Basel in 1987. There were eight of us craftsmen who participated in the fair under the banner of the *Académie Horlogère des Créateurs Indépendants* (AHCI). In the official information leaflet, the AHCI was introduced as a special exhibitor.

Furthermore, from 1987 to 1989, the grounds and an exhibition booth were made available to us free of charge. Afterwards, since apparently certain watchmaking firms took umbrage at that, we had to meet the cost of our booth like every other participant.

To explain the success of the AHCI, we need to go back a bit. In the 70s the appearance of quartz and the prophesied demise of mechanical watches had generated a crisis of conscience among watchmaking enthusiasts and inspired the founding of associations in several countries. Auction houses began taking a greater interest in contemporary mechanical watches. The attics of watch factories were stormed. Anything bearing a resemblance to a mechanical watch found its advocate.

Swiss watch manufacturers had abandoned the production of "special" watches—that is, those with complications—and horology schools no longer gave any instruction in the crafting of such movements. The AHCI came along at the right time then and could, without any competition, meet the demand for new models. Some watchmaking firms began to commission items from the members of the AHCI, since the Academy's craftsmen were capable of designing and producing these new creations which the firms themselves could no longer develop.

Collectors added to the mix. They allowed several of our members to thrive and eventually to found large watchmaking companies of their own. The first member to enjoy a meteoric rise at this time was Franck Muller, followed shortly by François-Paul Journe and many others. Some craftsmen kept in touch with the AHCI for a while without, however, becoming members. I can mention Renaud & Papi, Christophe Claret and Greubel & Forsey in particular.

From the start, the AHCI had rallied enthusiasts all over Europe. And then further afield. To my great delight, we enrolled our first Asian member, the Chinese Kiu Tai Yu, a knowledgeable collector and a peerless watchmaker. Kiu Tai Yu was an outstanding member and a sincere friend. Afterwards we also succeeded in gaining members from Japan as we do these days from China.

The original goal of the AHCI has thus in large part been met. But the survival of craftsmen is still in the balance, given the policies of the large manufacturing groups and their monopoly of the shops. We have nonetheless proved that, in tandem with large-scale manufacturing, craftsmen still have an important role in the eyes of knowledgeable connoisseurs. It was no accident that the Academy took the Special Jury Prize in the 2010 *Grand Prix d'Horlogerie de Genève*. Every year, several individual members have also won prizes.

At the Basel Fair in 1987 a jeweller from Milan passed on to me an order for a Spatial watch displaying the boot of Italy. This man, Anselmo Grimoldi, later became a very dear friend of mine. I made five such watches for him in 18-carat yellow gold, plus a sixth in platinum. Here I must also point out that I was the first—and still the only one—to have crafted a movement in platinum, a metal that is quite awkward to work with.

Italy in gold
and case in sapphire

Italy in platinum
and case in sapphire

VIII

The Real Turning Point

Since Piguet still had not brought out my flying tourbillon, at the Basel Fair in 1988, I presented my own in a Spatial watch, the first flying tourbillon in history in a wristwatch. Its appearance, at first sight overly simplistic, won it no respect. I left the fair empty-handed. Today—an ironic, but comforting fact—those in the business have understood that it is harder to make something simple than something complicated.

Flying Tourbillon, 1988

Approaching my 45th birthday, I was still not managing to live comfortably off my art. Feeling that I was misunderstood by the profession, I was losing confidence. It had been 12 years since I invented my Golden Bridge. I was killing myself at my work without being able to earn a living.

I was a hair's breadth away from dropping it all and giving up my craft. But I decided to play my last card. If that got me nowhere, I truly would give it all up.

The "Baladin": A Further Reward

And 1989 proved to be the year (at last!) of the Great Leap Forward, a revolution in my financial circumstances, my creative spirit and my way of understanding life. This is what happened. In December 1988, we were approaching the Vicenza watch fair, scheduled for early January 1989. Four years earlier, a financier had asked me to create an original mechanism for him. I had submitted several proposals to him, but he did not follow up on any of them. So I took one of them, made a working prototype and produced a watch. I called it the "Baladin" after the Italian *cantastorie* who wandered about cities telling stories. Its special feature was a display without hands, where you read the time by means of an aperture that wandered about the rim of the dial. The angular position of the aperture indicated the minute, while the number visible in the aperture indicated the hour. When the aperture arrived at the top of the dial, the number changed to indicate the following hour. The presentation was very simple and extremely readable, relying solely as it did on this alliance of two indicators. In Vicenza, I met Anselmo

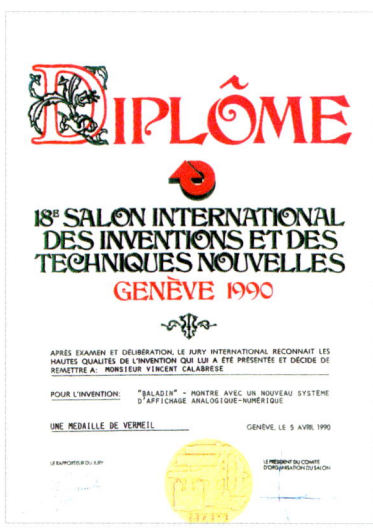 

Baladin, 1990

Grimoldi again. I told him that I was probably going to get out of the business. I nonetheless showed him my Baladin, the watch that represented my last hope. "That's fabulous!" he declared. He took the watch and it was immediately passed around the booths at the exhibition. A dozen manufacturers showed some interest in it, which eventually led to my selling it along with its patent to the Italian firm Pinko, which also commissioned me to make ten of my Esprit Spatial watches. This was unprecedented. I could not believe it. I had never made as much as 30,000 francs in a year, and now in one deal I had made several times that amount.

The Baladin won the silver medal at the International Exhibition of Inventions of Geneva in 1990.

Ten Flying Tourbillons in Two Years

Following this success, in February of that year I suggested to the editor of a magazine that we might offer a subscription for ten flying tourbillons, of the same type that I had exhibited at the Basel Fair in 1988. The watches would be differentiated by the client having their own choice of engraving. I also proposed that each edition of the fortnightly magazine would contain a two-page spread with updates on the progress of my work. This was a serious challenge I was undertaking. I was committing myself publicly to delivering my handcrafted tourbillons before the end of 1990—that is, in less than two years. I kept my promise.

The "Two Hands" Clock

At the end of March 1989, the Basel Fair came round as always. For the occasion, I had decided to create something unprecedented: a clock with a tourbillon called the "Two Hands". It took me three months to design and produce it, working day and night. I lost six kilos, but the clock made quite an impression. It consisted solely of two hands. The entire movement was housed inside the minute hand, at the end of which the tourbillon was positioned, an unheard-of configuration. Instead of a balance wheel, an essential part of every tourbillon, there was a pendulum that achieved "the impossible"—impossible, that is, in the eyes of tradition and current knowledge. In a stationary clock, the regulating organ is the pendulum. It can only function if the clock is immobile, whereas in mobile timepieces, such as a wristwatch, the pendulum is replaced by a balance wheel, which can function in any position. The icing on the cake: although in theory a pendulum cannot have an amplitude greater than a few degrees, that of my pendulum was greater than 260 degrees. That was a self-educated man's reply to the "experts" in the world of horology. Albert Einstein said, "Engineers can tell you why something doesn't work. Craftsmen cannot tell you why it does." However, I could. Let the experts figure it out. My Two Hands was a cry of rage in the face of the incomprehension that had greeted my Spatial watches.

I stopped making my Spatial watches, leaving them for better times when they would meet with greater acceptance.

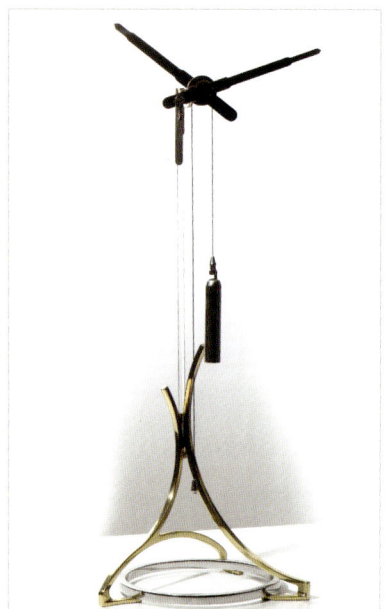 

Two Hands   Detail of the **Two Hands** model

Inventions, Creations, Incomprehension

I like to look at things in the cold light of day. That is my nature and one of my main strengths. I am a pragmatist. This is where I stood: the public, taken aback by the elimination of the dial and the use of sapphire for the case, considered my watches "outlandish". If I wanted to carry on in this fashion, I was going to have to set the bar higher: I could no longer rely on my Spatial watches but would have to follow the path upon which I had embarked with my Baladin. That is, I would have to stick with an ordinary movement, dial and case, but make them special by giving them an original interpretation.

Thus, my exhibitions at the fair were my Baladin and Two Hands. The Baladin charmed the media and a number of magazines ran articles about the creation of this watch. My 12-year nightmare seemed to be over. And yet I could not put it behind me. I was still haunted by the feeling that my message was misunderstood.

The Power of the Large Firms

In 1989, Patek Philippe was celebrating its 150th anniversary. For the occasion, the company produced a limited series of jump watches. The publicly announced price was 20,000 francs. They all sold for more than 30,000 before they were even available to the public. In the same year, Cartier brought out a similar watch. It did not enjoy the same success as Patek Philippe's, but it still sold very well. These facts (sadly) proved once again that people buy an image, not a product.

The "Commedia", First of the "Philosophique" Watches

In my own (theatrical) fashion, I had to decry this sort of speculation on the big names. Hence, the creation of my "Commedia". Given that it had a case and dial; it could be considered a "normal" watch in comparison to my Spatial watches. But it was an unusual sort of dial: it displayed a partly-open curtain on a stage. The hour could be read in the opening between the parted curtains. When the minute hand reached the top of the dial, the number indicating the hour instantly changed to the next. To reinforce the message, the Commedia ran on a quartz movement, even though such a movement is less powerful than a mechanical one. There were two reasons why I chose quartz. I wanted to keep the price below 1,000 francs as well as demonstrate that my work was technically superior to that of the large manufacturers.

I was fed up with watches whose prices were totally unrelated to their actual composition. Hence, the name "Commedia". It sold extremely well. I was seriously beginning to make a name for myself. Several articles about me were published in various countries.

I finally got some attention in Italy, which up to then had ignored me (no man is a prophet in his own country). Several journalists came to Basel to interview me. I will point out that at this time Italy was the Swiss watchmaking industry's second largest market.

Continuing on the theme of comedy, I recalled Dante Alighieri's divine one. That gave me the idea for a new model, a variation of the Commedia—the "Divina Commedia". The jump-hour display became a "jump-word" display, the words displayed adapted loosely from Dante's famous work: "Perdete ogni speranza, voi che create, l'arte pagante è solo alle . . .!" That is, "Abandon all hope, ye who create. Only the fashion of the moment pays." Sadly, my Divina Commedia was a failure. I only sold about ten of the watches.

The "Mona Lisa"

At the time, I carried on in this vein with another variation on the theme of the Commedia dell'Arte. In this version, the parted curtains revealed a series of images comprising a striptease act. I took advantage of the Vicenza Fair of 1991 to launch the "Mona Lisa". It was an immediate success. The watch was dedicated to women's liberation.

Commedia

As the hours go by, you see a woman removing her work clothes. At 11 o'clock she turns her back, completely undressed, freed from her sacrifices to the dictates of fashion. Finally, at noon you see a close-up of the famous face of the Mona Lisa and her enigmatic smile.

I had won my bet. Attracted by the erotic nature of the watch, in the end men were thrown by the Mona Lisa's derisive smile. As a criticism of the use of eroticism for commercial gain, I conveyed my message in a series of eleven erotic images followed by one that was purely cultural, the Mona Lisa. Echoing contemporary society, an original procedure of this sort can be termed "cultural". I sold several thousands of these watches, most of them to women, for the modest price of 1,200 francs.

Start of the Good Times

In the middle of the fair in Vicenza, we heard that the Americans had intervened in the Gulf War. After that, the exhibition was a disaster for many of the participants. Up to then Italy had been the second largest importer of Swiss watches. That came to a halt. Pinko shed its watch division. The brand, which had produced more than 2,000 Baladins, went bankrupt and my royalties were suspended.

Mona Lisa

Pinko's bankruptcy cost me a great deal of trouble. A dishonest competitor took advantage of this period of uncertainty by illegally acquiring a thousand Baladin movements and then producing watches under its own brand. They ignored my warnings, instead putting another thousand watches on the market. As a result, it was 1996 before I could buy back my rights from Pinko and take legal action. It was slow going, and I changed lawyers when I found that their own hands were not always clean. The third one got the culprit convicted in 2000.

1990 Blancpain Tourbillon

1991 enamelled Esprit version

IX

Cloud Nine

One day a manufacturer asked me to make a mechanism to indicate the power reserve. It was a good idea. You never knew how long a watch would run before you had to rewind it. An indicator on the dial would be very handy so that you could tell how much power you had left. The manufacturer's idea was especially tempting because the public was going over in droves to automatic watches.

Most mechanisms have enough power to run for about 40 hours. Other more complex ones can run for several days. When the power runs out, the watch stops, making it necessary to rewind it and reset the time. It's simple enough to conceive of a mechanism that would indicate the power reserve for a non-automatic watch, but it's much more difficult for an automatic, which is running down and rewinding at the same time, thus working in two different directions. Such a problem can be resolved only by means of a differential, but that solution involves you in one of the most difficult calculations in mathematics.

My partner already had such a mechanism, but when he showed it to me, I realized I could make a much better one. I made a prototype and offered it to him. Then, unexpectedly, he let me know, at this late date, that he feared the research costs he might encounter. He preferred to buy mechanisms that were already available. So once again, I found myself with an invention with which I could do nothing. Life is full of surprises however. A colleague of mine had also just received a commission from a group of watchmakers for a power reserve indicator. When he saw my prototype, he immediately halted his own research and suggested we enter a partnership. That was at the end of the 90s. The joint manufacture of some 15,000 of these mechanisms was launched straightaway.

The "Power", the First of the "Technique" Collection

I was on cloud nine. Now was the time to take up in earnest the manufacture of my "extraordinary watches with an ordinary movement" that I spoke of earlier. I decided to use my power reserve indicator. The first model would be called the "Power". It would comprise a round watchcase, identical to the one used for my tourbillons, with four attachments uniting three components to symbolize the 12 hours. (In my creations, I have always tried to devise a symbolic element to give the watch an extra dimension.) The case was in steel and the power reserve indicator was extremely efficient. Hence the name "Power".

Taking advantage of the expertise in jewellery making that I had acquired, I created a dial that was different from the usual in aesthetic and stylistic terms. First, the colour of the dial was a milk-white, similar to the colour obtained when silver is quenched in boiling sulphuric acid to eliminate the blackness caused by the heat of the soldering. This colour, dull silver with a hammered surface, had the advantage of not being reflective. It was ideal for reading the time. Secondly, the hands were fashioned of blued steel. They offered a striking contrast with the dull silver of the dial. Finally, I rounded off the overall effect with hour-markers appearing on a polished gold ring. The result was a multifaceted dial that was unprecedented at the time, a delightful aesthetic innovation inspired by the simplicity of earlier times and adapted to contemporary tastes. On the face of it, an ordinary watch—but enhanced by the lovely complication of the power reserve indicator, the thinnest ever produced, and with a reliable mechanism that I had ensured would not be subject to wear. I introduced the "Power" in 1992.

At the time, I was making a better living, but my overheads were high. Remembering the previous hard times, I decided to keep the price on this range of watches as low as possible. The first savings lay in always using the same case and movement. I finally adopted the

Power

Transworld

ETA2892 movement since in my experience it was the best performing on the market. Furthermore, this challenge was going to present me with another: since my case had no push-pieces, the implication was that resetting the time or correcting any of the indicators would have to be done by means of the winding-crown. Eventually I realized that it was by imposing such constraints on myself that I had been able to introduce some real innovations to the mechanisms themselves.

The "Transworld"

I then set about developing the "Transworld". As my clients were frequent travellers, they needed a dual-time watch, practical and functional, but different from existing timepieces. All watches that offered this function had, in my view, the same flaw: they insisted on displaying a traveller's home time in the center, which made reading local time impractical. So I decided to reverse the two displays.

The central hands would indicate a traveller's local time, whereas his home time would be indicated in small sub-dials. My objective was to allow the traveller to make full use of his watch, no matter where in the world he happened to be.

The fact is that not all adjacent time zones in the world are a full hour apart. In some cases, the difference is only a half-hour—and in certain Indian states even 20 minutes. So I chose to indicate home time by means of two sub-dials, one for the hours (with a 24-hour display) and one for the minutes. Furthermore, the manipulation of the watch needed to be kept as simple as possible: by means of the crown alone, which could be put in three different positions, it should be possible to rewind the watch, reset the hands of the dial and sub-dials and change the date.

I had found the answer! However, there was still the question of producing the mechanism. I did not achieve my goal at the first attempt. I had no examples to follow. I nonetheless finally succeeded by housing this complex mechanism, along with the power reserve indicator, in a case only 0.8 mm thicker than usual. I also managed to provide for all the necessary adjustments to be made using the winding-crown alone. As is common knowledge, every watch is manipulated by means of the winding-crown. Normally this can be put in two positions. By pushing it all the way in, you rewind the watch. By pulling it out, you can reset the time.

With watches that have additional functions, there is a third, middle position. Such is the case with the Transworld. It changes the date by turning the crown counter-clockwise and the central hands by turning it clockwise. That allows for the central hands to be moved to indicate the proper time zone without moving the hands in the sub-dials, which continue to indicate the home time. When the crown is pulled all the way out, it allows both the central hands and those of the sub-dials to be moved.

The "Daily" and the "Mobil"

I decided that from then on my watches would have to be of that same thickness. Included in this collection, which I called the "Techniques", was the "Daily", introduced in 1993, with power reserve indicator, dual-time function, date, the day of the week and day/night indicator; and the "Mobil", introduced in 1994, with a power reserve indicator, date and dual-time function, with a 24-hour display of the home time.

During this time, between 1991 and 1994, I also agreed to create six different mechanisms for six different manufacturers.

Daily

Mobil

The "Janus" for Universal

The most interesting mechanism, introduced in 1994, was cre‒ated for the centenary of the Maison Universal Genève. This was the "Janus", a double face watch, like Jaeger-LeCoultre's "Reverso". The reverse displayed only the hours and minutes, while the obverse had a jump-hour function, along with the minutes, day/night indicator and the date.

On December 13, 1993, I signed a contract for the acquisition of a patent and the design and production of 500 of these watches. On April 9, 1994, I delivered ten of them at an auction at Antiquorum, where they were sold at an average price of 35,000 francs. On April 14, 1994, the opening day of the Basel Fair, at five in the morning, I finished ten more for the exhibition. I was driven to the fair, which allowed me to appear on the booth fresh and rested. I leave it the reader to guess how well I held up.

As the icing on the cake, at the same time, Jaeger-LeCoultre introduced their first "Reverso", comprising complications on both sides. Being honest, I have to say that my Janus had more. It was also thinner. On the occasion of the centenary, Universal published a very nice book with my Janus on the cover. Sadly, and as usual, there was no mention of the fact that it was my creation.

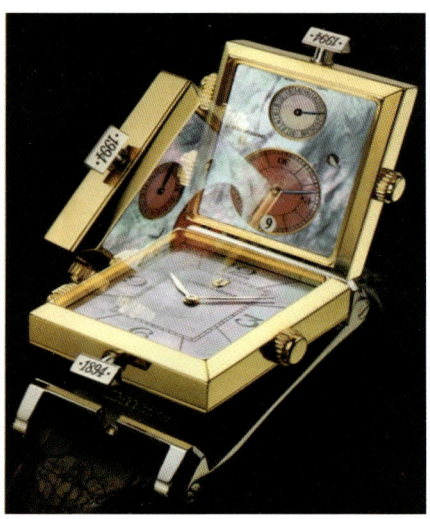

Janus for Universal Genève

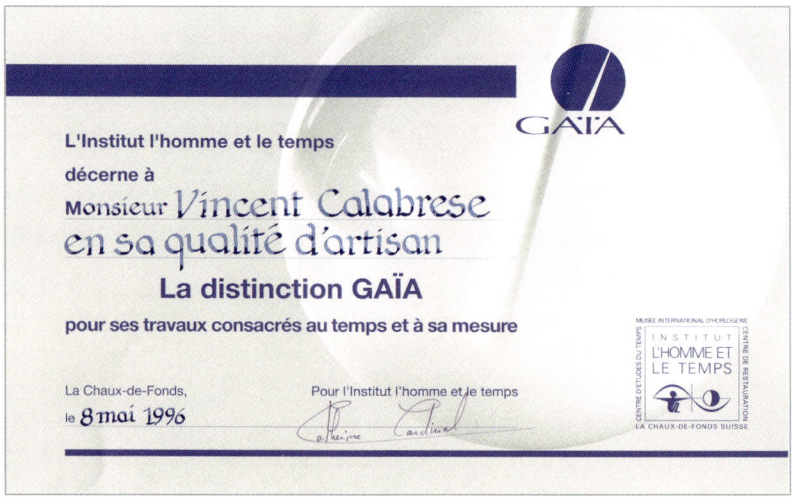

X

The "Ludique" Years

Night and Day

I presented Night and Day at the Basel Fair in 1995. While it could still be considered part of the Technique Collection due to its mechanism, the public saw it as one of the Ludique (playful) watches. It is nonetheless a Technique due to its traditional analog display. This complication was invented at the end of the 19th century following the adoption of Universal Time and has been patented any number of times. The function of the complication is to change the hour indicators at noon, with those of the hours before noon being distinct from those of the hours after. The solutions adopted were not, however, very reliable, and none has stood the test of time.

My implementation of this complication, by way of maintaining the distinction between the hours before and after noon, involved indicating the hours before noon with Roman numerals and those after with Arabic numerals. The dial had eleven apertures, ten for the display of the hours, and the eleventh, at 12 o'clock, for the date. From midnight to noon exactly, the apertures displayed the Roman numerals from I to XI.

Then at noon, they were replaced, instantaneously, by the Arabic numerals 13 to 23, with the same alternation starting up again at midnight. At 6 o'clock there was no aperture, instead an arc where a small hand indicated the power reserve.

Recognition in Japan

1995 was a significant year. The NHK, the Japanese national broadcasting company, offered certain of their "living treasures" an hour timeslot to report on a topic of their choice. The manga artist, Matsumoto, chose Swiss watchmaking. After a tour of the factories of the largest

Bun Pei

Pegasus

Tourbillon Phantom

Esprit Platinum

and best-known manufacturers, the program finished in my workshop, devoting a fair amount of time to my work. As a result, a distributor became interested in me, and I gained entry to the Japanese market by the main door. Through my watches, the Japanese discovered the AHCI, and many of my colleagues in turn got a foothold in this large market with its huge cultural curiosity, which also attracted great connoisseurs of *objets d'art.*

Japan became my principal market. I decided to take up my Spatial watches again in its honour, with the creation of various "Personnelles", including one displaying a Japanese monogram, as well as some of my "Symbolique" watches, such as the "Archipel japonais", the "Equus", the "Pegasus", and many of my "Esprit" and "Tourbillon" models, including the newly-created "Regulus".

Re-acquisition of my Baladin Patent and the Prix Gaia

In 1996, happy that I had finally been able to repurchase my Baladin patent from Pinko, and seeing the interest shown in Night and Day, I continued working along these lines and began producing my "Ludique" collection. In the same year, as a reward for my contribution to the Swiss watchmaking industry, I won the Gaïa Prize. Created in 1993 by the *Musée international d'horlogerie de La Chaux-de-Fonds*, this award is designed to recognize those who have brought about a significant development in the measurement of time.

The Technique "52"

1996 also saw my 52nd birthday, and by way of a present, I treated myself to the jewel of my "Technique" collection, the newly created "52". Using the same case (and hence the same volume), I managed to shave off 0.2 mm; and within an additional 1 mm, I created a mechanism driving dual time, power reserve, date, day of the week, day/night, week and month

"52" watch

functions, all of which could as usual be modified via the winding stem. Clearly, handling of this watch was a complex affair and an owner's manual was included with the purchase.

The "Horus"

1998 witnessed the creation of the "Horus", the third watch in the Ludique collection, featuring a plain dial without indexes, but with a large seconds hand in the centre. Along the rim, positioned like a satellite, a sub-dial displays the 12 hour indexes with a small hand in its centre to indicate them. It is the position of the sub-dial that indicates the minutes, for like a true satellite, it makes a complete revolution every hour. Its distinctive feature is that it remains vertical throughout its revolution around the dial. This would not be something commonly seen in nature. For example, consider a "6" at the edge of a circle: as the circle turns, the "6" will turn with it, becoming a "9" at the 180° mark and again becoming a "6" at 360°. In the case of the Horus, the "6" will remain a "6" throughout its journey around the rim.

Horus Steel   Horus Gold

New Workshop, Reorganization and Prosperity

The 1990s were a time of great change, both in financial and social terms. In 1985, I left my shop in Montchoisi and took another one on the Boulevard de Grancy.

It was a better location, near the train station, and at a lower rent. It was not terribly big, giving me only 19 square metres for my workshop and a small area where I could greet my clients. Eventually I abandoned the commercial side of the business and devoted the entire space to my workshop. It was not easy to move around amongst all the second-hand machinery that I had at last managed to procure.

A significant event in 1992 was the return of my younger daughter from Australia, where she had gone to learn the language. She now spoke English perfectly, and I suggested that she should come and work for me to handle my relations with my clients and the media and above all the administrative tasks, so that I could focus entirely on the creation and manufacture of my watches.

At the end of 1993, I learned that in the upper basement of a neighbouring building there was a two-room office and a study, altogether measuring 69 square metres, for rent. The rent was five times higher, but as my business was beginning to thrive, I was feeling more and more cramped in my little shop. When I visited the premises, I was pleasantly surprised to find that they were well lit with windows looking out on the "Ficelle", Lausanne's first metro line. As Lausanne is built on a series of hills, there are many buildings with the ground floor on one side, while the lower floors are on the other with their windows looking out on the street. I was taken with it straightaway, especially since an apartment was just opening up in the same building. Two years later, I rented the adjacent premises and thus had a total space of 195 square metres available to me.

Everything was going well! I was receiving all sorts of commissions, from rather modest mechanisms to complete watches. In 1994, I had to hire one watchmaker and in 1995 a second. My workshop was filled with new light. My first clients entered with a look of pity, saddened to see me relegated to the basement, until they discovered my spacious workshop and sunlight streaming in through the large windows. They came from all over the world. Furthermore, I no longer had any trouble placing my watches with distributors or in shops. And the media were reporting more and more freely on my creations.

I have never sold any watches through my own efforts. People have always come to me to buy them. That is very lucky for me, since I have a horror of the promotion side of a business, engaging in which would have indicated that my watches could not sell themselves. For this reason, the commissions I received put my mind at ease, as did my big Japanese clients. They came without any need for me to encourage them. The business that I did with the Japanese encouraged me to hire my older daughter as well, along with other watchmakers. In the end, there were eight of us altogether. That said, what it meant to be an employer was not always an easy thing for me. Though I hoped I might be mentoring disciples, I was in fact dealing only with wage earners. Creativity is not given to everybody.

Continuing Crisis and New Commissions

Japan had become so important to me—almost 80% of my turnover—that when the crisis of 1999 struck, I had to let go my entire staff except for my younger daughter. Other markets did not offer me any hope. As a result, I went back to the commissions I had left on hold since 1995.

That which I undertook for the Enigma was thoroughly satisfying. In the early 90s, Enigma had brought out a highly original watch with jump-hour function that was very successful. Unfortunately, the movement they utilized was too complicated and intricate and did not allow continued production of the watch. They then asked me to find a solution that would get them out of this hole. I accepted the challenge, and my efforts allowed them to sell several thousand more of these watches without any technical problems.

XI

The New Millennium

The "Vingt-Cent" and the "Two Times"

While I was doing my training courses in 1971, I was impressed with the phenomenon of light refraction. Later, during my time in Crans-Montana, my dealings with my clients got me interested in gemmology, particularly diamonds. When I learned that a brilliant cut diamond is the most beautiful, I recalled the phenomenon of refraction. Nearly 30 years later, I decided to employ it in one of my own creations.

This was a variation of my Baladin. The mechanism in question, since it had no hands, enabled the creation of features never before attempted. With a display limited to a window at the rim of the dial, the center of the watch remained free. Then it was a question of providing for a synthetic white sapphire crystal with a blind hole in the center. The seconds hand would make its appearance inside this blind hole.

The overall objective was to allow the time to be read as usual by looking at the watch straight on, but as you tilted it bit by bit, the display would gradually change and finally disappear altogether, creating a special optical effect. The result was the apparent disappearance of the inner part of the face, with only the small, central dial remaining visible. The explanation of the effect lies in refraction: light easily enters the blind hole when it hits it directly; but when, on the other hand, it hits it from the side, it is blocked by the cylindrical wall of the blind hole, is diffused differently and so creates the expected optical effect.

The watch was launched in the year 2000, and I called it the "Vingt-Cent"—that is, "twenty hundreds" or "two thousand"—and the fact that "Vingt-Cent" is pronounced exactly like "Vincent" made the name even more appropriate, since it showed that I was staking my claim to the watch. A story from my childhood also provided a link between the watch and the new millennium. When I was little, we believed that 2000 would be the year of Martians and interstellar travel. Travelling through space entails certain dangers, however, and black holes, which swallow up everything in particular. My "Vingt-Cent" was an allusion to these black holes. I introduced this Ludique model at the Basel Fair in 2000, along with another new creation from the Technique range, the "Two Times". Housed in a new case, though of the usual size, this watch displayed the hours and minutes twice, as well as the day and a large date in a double window. I produced it as a limited series of 20 watches, that is, one for each century. These two watches aroused a great deal of interest, and I received a number of commissions for them.

Vingt-Cent

2 T

Commissions from Pamp and Bell & Ross

The first commission came from Pamp. This supplier of gold, known throughout the world, had discovered a very old rectangular watch with jump-hour display, and wanted to produce a replica, using a modern movement, as a 100-piece limited series. It met with great success, so Pamp produced another series of 150 of these watches.

The second commission came from Bell & Ross. This new firm also wanted to produce a jump-display watch. I took to the idea immediately and created a special model to be produced as a series of 100, with a jump-hour display and a power reserve indicator. This successful venture also led to the production of two further series of 100 watches each.

Commission from Goldpfeil

I accepted a third commission from Goldpfeil despite the fact that I had more than enough work to do. At the time, this world leader in high-quality leather goods had decided to make its grand entry into watchmaking and had contacted the AHCI about the production of exclusive watches. I was particularly interested in this proposal since it would involve the AHCI—seven of us, as it turned out. We were given a year to come up with a one-of-a-kind watch, as well as another intended as a limited series of 100. This was in September 2000. Besides my work for my own clients, I was required to complete and ensure delivery on these three commissions for the Basel Fair in the spring of 2001.

The Vingt-Cent, with its Wandering Hours display, did not figure among these other commissions, which included two jump-watches. Why not now produce two other jump-watches for Goldpfeil? I should point out that these four watches were to feature both a unique look and mechanism. It seemed like madness! I was fully up to the challenge however. For the one-of-a-kind watch I was to create for Goldpfeil, I came up with a jump-hour display in a window in the centre of the dial, with the minutes indicated by the entire dial as it rotated around the hour display. This watch would later evolve into the "Sun-Tral", the first watch in the world with a mechanism of this kind.

I was not involved in selling the watches produced for Pamp or Bell & Ross. Once they were delivered I was relieved of all commercial duties, except of course for after-sales service. That didn't bother me however.

It is one of the advantages of working on your own. I do not deliver my timepieces until I am completely satisfied as to the quality of the work and my internal controls. This means there are very few returns. Nevertheless, our contract with Golfpfeil provided for our full collaboration, so that the AHCI craftsmen became responsible for all the marketing, including promotional tours all over the world, while at the same time meeting delivery deadlines. The watches were to be exhibited at the

Pamp   Bell & Ross

Basel Fair in 2001, the collection having been pre-sold in various shops in a number of countries and a huge advertising campaign having been conducted in several languages.

Then two disasters struck, one after the other: the stock market crash in May and the September 11 attacks. Shops cancelled many orders immediately. There was also an auction of our seven special watches scheduled for November 20th 2002. My colleagues' watches were offered at a starting price of 50,000 francs. Bidding on my own offering, the precursor of the Sun-Tral, for some unknown reason, only started at 5,000 francs—except that there were no bids on it. Of the seven watches, only one, Frank Jutzi's, found a buyer. Goldpfeil finally went bankrupt. I nonetheless had a bit of luck there, since Golfpfeil had sold my series of 100 as well as 80 others with special functions. I recovered my intellectual property rights just in time. This was a serious blow for my AHCI colleagues however.

Goldpfeil Series   Goldpfeil Special

Portrait of Vincent Calabrese that appeared in a Bell & Ross catalogue.

XII

The Creative Spirit

Why I Create

My need to create, unconscious until I was 31, rests, as I have already said, on my need to give a meaning to my life, to leave a trace. I never imagined that this would be achieved through watchmaking. It was not until I had created my first watch—with the resulting tears—that I found the means to make an impression. As with any athlete, artist or researcher, I had to undergo intensive training in order to be able to envision new creations constantly, each one more high performance and original than the last. Isolated and harassed by financial worries, I was determined to carve out my own path. Watchmaking had gotten its hooks into me.

If, despite my labours, I would never succeed in mastering time, I strengthened myself, in my own way, by contributing to the effort to measure it. Just as one turns to a mirror to see oneself, I needed competition in order to take my own measure. The organization of the AHCI was in this regard a factor that helped me break the shackles of the large firms as well as bring my colleagues into the game in a sort of healthy competition. I wanted to engage in a fair battle, without ever yielding to the temptation to harm them or take advantage of them in any way. Thus, I got a glimpse of my own unique path. Not as a competition with rivals, but as a competition with myself. All that remained to me was the will to remain on the straight and narrow path, faithful to my own truth.

How I Create

My main strength lies in pure analysis. I have a need to examine the underlying design of things, the potential flaws in existing creations, the possibilities for improvement—in a word, I need to invent. I cannot be free from a desire to surprise people, to shun the well-beaten path, to go further, to go elsewhere, to be consistently demanding in respect to myself but not to others. I like to approach problems from a completely different angle, to imagine, to evaluate the emotion that I am trying to arouse, so that I might subsequently go in search of a technical solution that will enable me to arouse it.

Other Creators

I have great respect for many other creators. Some of them are very talented, and capable of fashioning complications that I would not undertake. That is because I do not have the patience to spend an entire year on a single mechanism. I will not mention any of my contemporaries, since that would be unfair to the others.

However, referring to earlier times, I will gladly mention Pierre Le Roy (1717-1785), inventor of the detent escapement, who remains a model of restraint and of tried and tested honesty. I will also mention John Harrison (1693-1776), who was never forgiven for being a self-educated man and who nonetheless enjoyed unparalleled success in chronometry. Both are men with whom I feel a strong affinity—though I would not go so far as to idolize them—something that is very fashionable these days, with the prevailing unthinking, pathetic need for gods to worship.

XIII

Nouvelle Horlogerie Calabrese

In 2003, despite my breakthrough in the world of watchmaking, things got complicated. Until then, the success of my venture had been due to my younger daughter, Tina, who relieved me of all the administrative work. Sadly, a sudden death affected her directly and signalled the end of these wonderful years. My daughter needed a change, and it would be impossible for me to carry on my own. To manage my growing business, which could involve up to 800 watches a year, all of which I produced myself, I took on a partner with whom I formed a limited company.

To mark the transition from small-scale production, which is how I would have qualified my business up to then, to a more industrial enterprise, and not to mislead my clients, we called our limited company the "Nouvelle Horlogerie Calabrese" (NHC). Initially the only model we introduced under this new brand was the Vingt-Cent, though it had a larger case and a completely new design.

In 2004, we exhibited our first models in Basel, with our usual success. I had created four new watches, the "Ottica", a reprise of the Vingt-Cent, the "Beauty-Fuel", the "Central Power" and the "Analogica", on the last of which a patent had been taken out. They immediately enjoyed huge success. In our first year, our turnover was well above a million francs.

The next year we brought out our "Ora" collection, which featured a jump-hour function. Jump-hour display involves a window through which you can see an image or read a message. Every hour the dial jumps 30 degrees to reveal another image or message. Then a big problem arose: my clients, accustomed to my prolific creativity, were expecting me to bring out the same number of new watches every year.

Ora Collection

Fine! I still had to make them however! It is no surprise that my partner also wanted to go this route but he wasn't the one doing all the work!

Ottica Wandering Hours display and sapphire crystal with optical effect. Disc with logo as a seconds hand.

Analogica Wandering Hours display and date. Disc with logo as a seconds hand.

Central-Power Classic display of the hours and minutes, windows for the date and day/night indicator. Power reserve indicator with 170° display.

Beauty-Fuel Dual-time display. Day of the week and day/night indicator, date, power reserve indicator with 180° display.

Furthermore, there was the risk of disappointing my regular customers if they were expecting new additions every year. The two of us ended up in total disagreement. To escape from the impasse, I suggested that I could buy his share of the business. This done, I discovered that the corporation's finances were shaky. As the shortfall was too large, I was unable to steady the ship and the venture ended in bankruptcy in 2006.

Fl'Ora 40 mm

TIME AND A LIFETIME

Col'Ora 32 mm

Fly'Ora 36 mm

Fl'Ora 36 mm

Avvent'Ora 40 mm

XIV

Hard Times

The period from 2006 to 2008 was another in which I had to attempt to survive. It appears that one cannot be a creator and salesperson at the same time. I had not lost my creative impulse, but the commercial side of things was pure drudgery. Worn out, I sold the rights to my mechanisms to Cartier. I needed a partner who would produce my creations, leaving me free to pursue my research.

Initial Contact with Marc Hayek and Blancpain

For a number of years I had enjoyed very good relations with Marc Hayek, a kind and very courteous man. Having taken up the reins at Blancpain in 2001, he had learned that I was the inventor of the eight-day flying tourbillon and wanted to meet me. Sometime later, he paid me a visit in my workshop and asked me to consider coming up with something new for Blancpain. Still struggling with my problems with the NHC, I put him off. I did not think it wise to get into some new venture. Now that I had taken full control of the NHC and brought the staff back on board, I urgently needed to restore the company's liquidity. As was usual in hard times, I accepted a number of commissions. The first came from Sellita, the largest firm concerned with the assembly of movements. They manufactured hundreds of thousands of them each year for ETA, who was happy to take them since they were of such high quality.

The Commission from Sellita

The elder Hayek, CEO of Swatch, and hence of ETA, flattering himself that it was his mission to save the Swiss watch industry, had the idea of breaking the supply chain of ETA's movements. This plan was not aimed at the *large* firms. They did not need saving. The *small* firms understood very well that such a move would threaten them. I have to say that I shared their opinion. Sellita reacted straightaway, as they were aware of the dire need to produce a movement that was entirely their own in order to free themselves of their dependence on ETA. I accepted a small commission to create a movement for them. Today Sellita is Swatch's main competitor in both the manufacture and assembly of movements.

The Commission from Vuitton

My hunger to create and above all my urgent need for liquidity obliged me to accept other commissions, and one that came from Vuitton in particular. They were intrigued by the possibility of producing a watch called the "Tambour Mystérieuse". The contract was signed—except that in the meantime, as I had informed them, I was negotiating with a large firm concerning another big project. The "Tambour" unfortunately remained mysterious and never saw the light of day.

The Commission for the Blancpain Carrousel

As a reply to unfounded criticism that some experts had directed at our tourbillon, Blancpain commissioned me to devise a karussel that was revolutionary enough in nature to put the tourbillon in the shade. This was a bold idea, but Blancpain insisted. They wanted to be the first to produce a wristwatch with a karussel. The truth is that there had been talk about this project for two or three years, but it seemed impracticable to me. I pondered it at length, trying to come up with an answer. Then one fine morning in August 2006, as often happened, when I woke up, I had it! Simple but effective, and more than I had hoped for.

Without going into the technical details, here is why this innovation was so important: in 1801 Abraham-Louis Breguet, a Swiss-born watchmaker living in Paris, and a man who was received in every royal court in Europe, invented a mechanism with a rotating escapement that he called a "tourbillon". Subsequently, in 1892, Bahne Bonniksen, Danish-born but living in London, invented a similar mechanism, though the design was different. The two mechanisms were almost equally efficient as regards keeping time. However, the profession did not like the idea of Bonniksen encroaching on Breguet's territory. Various stories made the rounds, allegedly confirming that the karussel could not match the tourbillon. The Blancpain Carrousel (as the brand decided to call it) proved that on the contrary, Bonniksen's invention was technically more efficient than Breguet's.

At the Basel Fair in 2007, Blancpain exhibited the prototype of the karussel developed from my work and that had features never seen before: this karussel, which rotates once per minute, puts the balance wheel at the centre of the system and can function for more than five days. This was an important moment in the history of horology. I had come up with another fine innovation 25 years after my flying tourbillon. It was enough to silence a century of carping and to re-write the book on watchmaking—and, inevitably, arouse the jealousy of certain colleagues. A team was mobilised to manufacture this karussel as quickly as possible and get it on the market. It also led me to ask a question: was I really going to continue tearing my hair out when I could possibly go into collaboration with Blancpain?

Sale of Vica Sàrl and Collaboration with Blancpain

I saw Marc Hayek in his office at the Basel Fair and explained to him my desire to find a permanent and reliable partner who would enable me to devote myself solely to my creative work. A group like Swatch, with

Blancpain One-Minute Flying Carrousel mechanism

its many different brands, would be ideal. He was extremely interested in and in fact excited by, this idea. We mulled it over for more than an hour. He saw the possibilities for Swatch that such cooperation could bring.

Given the 18 brands the group had on the market, every one of my watches would find its niche. Once again, things were going my way.

We followed up on the idea immediately, but it proved to be a slow process. The agreement was finalized in March 2008, after a year of great worry for me. To observe the proprieties, I had halted deliveries to my clients and withdrawn from the last commission still in effect, that which I had got from Vuitton. I had informed Swatch, and they had no objection to my terminating it. Vuitton, for their part, did not want to press the matter, though they did not terminate the contract. Which is to say that, once other engineers completed the project, I should have started receiving royalties—though that did not happen.

Four Happy Years at Blancpain and a Mystery

My four years at Blancpain were among the happiest of my life. They had faith that I still had something to offer them despite the fact that I was 64 years old. They gave me complete freedom to pursue my creative work, along with the support of all the company's resources. This was a dream come true. Thus, still based in my workshop in Lausanne, assisted by two of the Blancpain staff, I could focus on pure research! We got along perfectly well, no one more surprised than I, the lone wolf, to discover such harmony. The only blot on the landscape was the deterioration of my good relationship with Marc Hayek, which was suddenly broken off for some reason I could never discover.

This was nonetheless an extremely creative period for me: nine patents in four years, without counting innovations that were not patented. Among those patented, the two that I was most proud of were those taken out on improvements to the Karussel and to the tourbillon, the latter of which, sadly, Blancpain did not follow up on. I had thus more than lived up to my side of the bargain. And if I have surpassed the innovators of earlier times, there's no doubt that those to come will surpass me. It works the same way in sports.

Baselworld, 2008. My last exhibition before Blancpain. Fondue dish with spaghetti, two coffee cups, the Gulf of Naples in the background, and my watches drying in the breeze, like the laundry in the photo of Naples in the background.

Yet it was some time before I realized that Blancpain, oddly, never tried to put any of my innovations on the market. I took out my patents one after the other. But it seemed the firm always had other products they needed to launch more urgently. I was champing at the bit: it was

becoming quite exasperating. I could not understand why, after such a fine start and excellent reactions from the media and watch connoisseurs alike, I was cut off from contact with the outside.

This went on for two years. Then the company decided that the rent on my workshop was too high. I was transferred to headquarters in Paudex. At first, I tried to see the move in a positive light, thinking it was a good sign that I was being put at the centre of operations. Perhaps it would be a chance to restore dialogue, to my advantage, with the CEO. It did not turn out that way. I held on for another year, working part-time, to finish all my projects and prepare to take retirement.

The falling-out with Marc Hayek remains a mystery—as was the freeze on my latest innovations and the failure to take advantage of earlier ones. I had no choice but to hand in my resignation. That decision surprised no one. They were all aware of the situation—which could properly be called surreal: the company was paying me without making any use of my work! I had to wonder if the reason was only to prevent someone else from doing so. My contract stipulated certain firms for which I was forbidden to work for a year after my departure. I honoured that provision and tried to forget about the creation and manufacture of watches.

2009

XV

New Life

In my passion for my profession, I fear I neglected my family. Firstly, my two daughters, who acquainted me with the happiness of being a father and persuaded me to support the fight for gender equality. I thought I was giving them the time they needed. I was wrong. They no doubt felt neglected. Even if they have forgiven me, I still bear the sadness of having created the feeling of abandonment in them.

Neither was I able to bid farewell to my father. He died in Naples in the 1980 earthquake. I only saw his lifeless body in a hall heaped with many others. The grief of the survivors put my own into perspective. Seeing a child who was the sole survivor of a family of six made me understand the extent of the disaster.

My mother's death in February 2000 filled me with remorse for not having given her enough time, either. I was there for her, just as I was for my daughters—but they had to deliberately call for my attention.

After 37 years of marriage my wife asked for a divorce, something I never would have considered, given the weight of Italian tradition. Our marriage had not been successful, despite the compromises we had made. We were simply too far apart, and nothing had changed with time, but our family life was the most important thing in my eyes. My daughters needed both a father and a mother. I did not have the courage to try selfishly to make a new life for myself, but my wife's insistence won and we got divorced. From this point of view, 2001 was the year of my rebirth.

My partner, Évelyne, came into my life in 2003. She was, and remains, the happiest discovery I have ever been granted. She has been my guardian angel, my support throughout these recent turbulent years, my rest for the warrior in moments of doubt and weariness, but even more, my source of energy and an oasis of happiness.

New Start

It was now December 2011. I left Blancpain and at the age of 68, I decided to attempt to take a more relaxed approach to my affairs—

not that there was any question of living like a typical retired person. I needed to find some sort of compromise. Our circumstances in Lausanne did not encourage change. We were still living in the three-room apartment in the same building as my workshop. That did not help me to put a damper on my passion for my work. I needed to burn my bridges. Despite everything, we wanted to remain on the shores of Lake Geneva, and so we moved to Morges. I had lived there in the past. In fact, my daughter Marina was born there. This pretty little town was even more charming than before and had lost nothing in terms of quality of life. We wanted some peaceful corner to live in where we could enjoy the beauty of the countryside.

We quickly found an apartment such as I never would have dreamed of as a child—with a lovely view of the lake, the Alps and the Mont Blanc! My life has been marked by three mountains: Vesuvius, at the foot of which I was born; Fujiyama, witness to my success in Japan; and Mont Blanc, the dream setting for my old age. Now all I needed was to find a little workshop where I could provide after-sales service for the watches sold during my time of independence. I tried three before finding one that suited me, one with the space that allowed me to undertake various projects with my friend, Jean Kazès.

Jean Kazès

I had had the pleasure of seeing my work recognized during my lifetime—but that made me think of my friend and colleague, Jean Kazès. A creator of sculptural clocks of exceptional beauty and originality, he is a truly independent craftsman. Despite being quite limited, his production could not find any buyers. Furthermore, since the manufacture of his clocks required expertise in so many different areas, it was going to be impossible for anyone, even a very skilled clockmaker, to carry on his work. The principal asset of his creations is their aesthetic aspect in that they are true sculptures. He has often been compared to Tinguely. In Jean's work, one notes the exquisite gilding, the polished steel. His works are made for the sitting room, not for the garden. He is a very discreet and modest, in fact humble, person who cannot endure any sort of fuss. If you tell him his works are beautiful, he murmurs, "Yes," almost timidly. Since I had not been able to perpetuate my own name, I made it my objective to perpetuate his.

In 1988, when we were exhibiting our works together for the first time at the AHCI stand, we were intrigued by each other's creations and the freedom and daring they expressed. After this, we exhibited our works together every year. A spirit of independence created a bond between us. Even if my success had led me to stop exhibiting my works with the AHCI, from 1985 to 2008 I supported him by continuing to exhibit his clocks at my stand at the Basel Fair. Was there not a danger that he, too, might be forgotten? After all, he was in his 80s.

I insisted on buying the rights to his works and registering a trademark on his brand name (something he had never done) in order to protect it. We signed an agreement, stipulating that he had the right to change his mind for the next five years. We are still very close and quite happy with the step we took. It was the confirmation of a beautiful friendship that has lasted 30 years since the time he joined the AHCI. He has not changed in any way and still demonstrates the same spirit of independence in his work. I will only be concerned about him when he decides to stop producing.

I began learning about how he went about creating his clocks. I set up a website about them, in addition showing some of his models in 3D to give a viewer an idea of their size. This did not help much with his sales however, although the number of visits to the site is constantly growing. I also organized some exhibitions of his clocks. Together we created a clock inspired by the classic grandfather clock, preserving the movement in its entirety as well as the original rectangular shape. We also utilised the chains that are typical of Jean Kazès, thereby giving the end product a modern look—and all this at a modest price so as to help his work gain a wider audience.

Golden Bridge 1980

XVI

Corum

In 2012, against all expectations, a contact at Corum suggested that I renew my ties with the company. This surprised me. From the very start, the owners of the firm, far from being grateful for the watch that I had allowed them produce, had not shown me the least respect. At the end of the 90s, facing imminent bankruptcy, they were forced to hand over their operations to Severin Wundermann. This transaction, from my point of view, was a case of out of the frying pan into the fire. Not that I could express my opinion about the new owner, who in any case had never shown any interest in me. That did not prevent him, however, from producing a new movement for the watch without informing me. In the mid-2000s, Wundermann, by then in poor health, created a foundation to which he handed over his authority as director, also naming a new CEO to take over the management of Corum, a young wolf who had cut his teeth at Panerai—Antonio Calce.

Ambassador for Corum

Antonio Calce had a great deal of trouble getting Corum back on its feet. The firm had neglected its flagship product, the Golden Bridge, in order to focus on more mainstream products. The foundation management, based in the U.S., took their time before demonstrating full confidence in Calce and allowing him to bring the Golden Bridge to the forefront.

Once given the chance, he decided to put an end to the stupid antagonism between us—and to use the inventor of the watch to best advantage, since I was still active and furthermore, a free agent, now that I had no further obligations toward Swatch. Calce proposed making me an ambassador for Corum.

Given that the Golden Bridge was still selling quite well 40 years after its creation, and in fact represented 60% of their sales, it was in our mutual interest to mend our fences. So in January 2013, we made a press announcement about the signing of a contract that restored ties between us by naming me as Corum's ambassador.

High Treason

This good news was followed, however, by some bad news. Blancpain reproached me for my new role with Corum. I replied that I was not working for Corum. I was simply their ambassador. Corum was not listed as one of the firms I was forbidden to work for once my collaboration with them was at an end, and in any case the year's ban stipulated by the contract was now over.

So what could be the explanation behind Blancpain's reaction, given that I had observed our contract to the letter? Many people agreed with me that their takeover of my business was intended simply to remove it from the market. A lawyer friend of mine assured me that Blancpain had no right to act as they had. Furthermore, he explained that it had been five years since I had sold my trademark and a trademark that has not been exploited for that length of time passes automatically into the public domain. This was my chance to recover my trademark. I was galvanized into action by the hope of bringing my work out into the light of day once again. Thus, I re-registered my trademark and Blancpain was urged to renounce it. To this day, we have not been able to settle the matter between us.

The Commission from L'Épée

At this juncture, a friend of mine, Arnaud Nicolas, CEO of L'Épée, contacted me. He wanted something truly special to mark the upcoming centenary of his firm. Recalling my "Two Hands", the clock that I had never put on the market, I proposed developing it. I envisioned a revolving movement, that is, a tourbillon, which would complete a revolution every hour, since it would be housed in the minutes hand. The escapement, placed at the end of the minutes hand, would be fashioned as a classic tourbillon, revolving once a minute, and, to top it all off, would have a 40-day power reserve.

I exhibited two special models of the clock at the Basel Fair, their housing consisting of lacquered spheres, one in titanium and one in brass. So there I was again, caught up in the fever of creativity! Now, foreseeing the recovery of my old trademark, I went back to work to prepare my return to the market.

Two Hands: double Tourbillon l'Epée, of which just two were created.

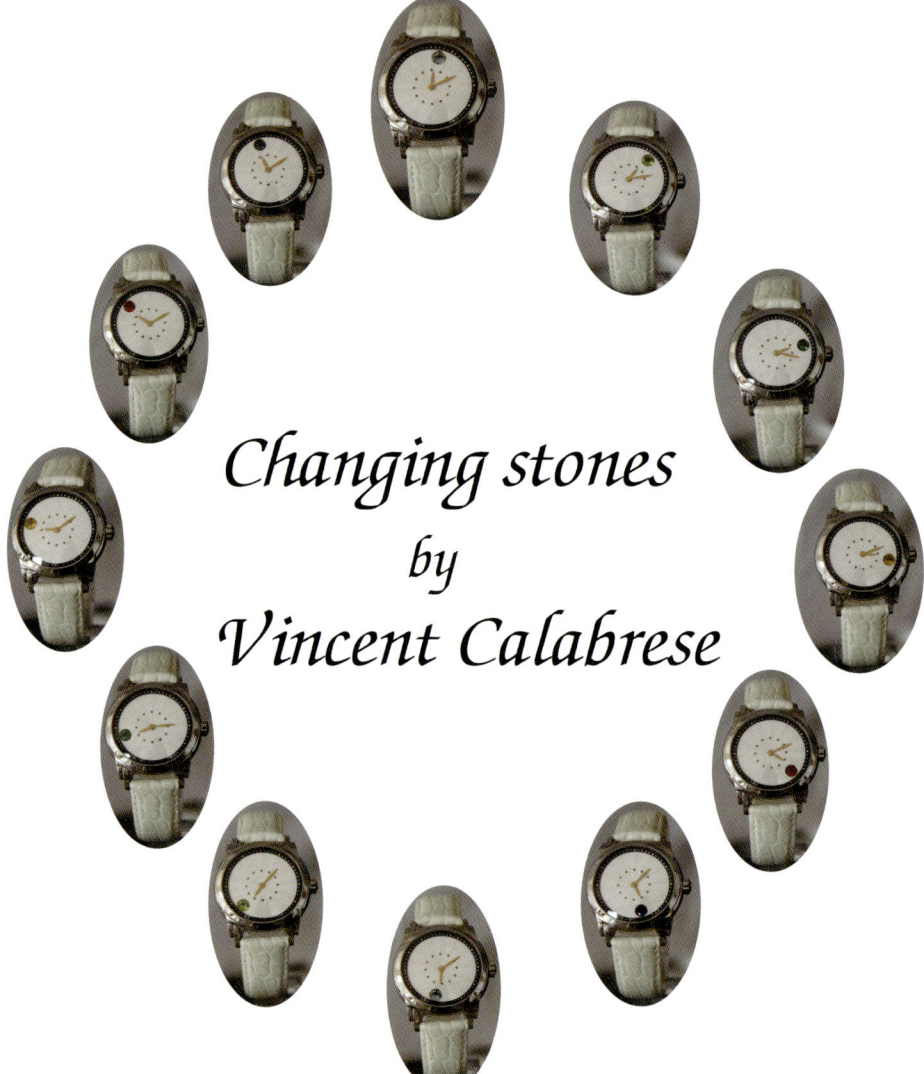

XVII
2015

The year 2015 was one of my best. It saw the celebration of the 30th anniversary of the *Académie des Horlogers et des Créateurs Indépendants* (AHCI) and the renewal of my creative work in collaboration with Shinji Himeno, a Japanese artist. My return to "business" was well-received. I now have more work than I can handle.

Thirty years! Thirty years of unshaken faith to preserve the original spirit of the Academy, to preserve the original purpose of the most beautiful of my creations. It has not always been easy. Some of our members have talent and a sense of ethics, while some have one or the other but not both. Those who had neither did not last long. The Academy is now grown up. During the period that I managed its financial affairs, I helped prepare its 'christening' with a major first exhibition at the *Musée d'horlogerie du Locle* and subsequently its appearance at the Basel Fair. Our initial goal was to exhibit our works at a joint stand so as to reduce the cost and make a greater impact. Once this goal was achieved after ten years under my management, it was time for me to withdraw.

For the tenth anniversary of the ACHI, I organized an exhibition in its birthplace, the *Musée d'horlogerie du Locle*. For its 20th, I took on the project of an historical brochure.

For its 30th, in 2015, I sought the support of the *Musée international d'horlogerie de La Chaux-de-Fonds*, which offered us an opportunity to hold an exhibition during the summer months.

I will not hesitate to emphasize how much the watch industry owes the AHCI. Our creations, the many commissions we have fulfilled for various firms, our members who have gone on to become important manufacturers in their own right, and above all, the spirit of emulation we have induced have made us a significant factor in the rebirth of the watch industry.

The stones

At Baselworld 2015 I introduced a prototype of a watch that I had wanted to create for a long time. While still utilizing the movement created for the Ora collection, I wanted to produce a new model for men and women who were more interested in a watch as a piece of jewelry than for its purely technical aspect. It would be a jump watch that would display, with every new hour, a different gemstone. My chief concern—and the chief difficulty here—would be the size of the stones rather than the images displayed, as was the case in the original Ora collection. Thus I decided to create this watch in two different sizes—36 mm for women and 40 mm for men. I was still trying to push the limits of the possible, and for a case of 40 mm the diameter of the stones could not exceed 5.20 mm, or a half carat. The large size of the stones only left me the possibility of employing a small quartz movement, with an autonomy of a year and a half. Or else—the height of versatility—to manufacture a bespoke mechanical movement. These seemingly endless variables obliged me to present only a single demonstration prototype and to limit production of the watch to special orders in accordance with a given client's preferences as regards the stones and the movement.

I demonstrated the feasibility of this model by exhibiting two perfectly functional watches at Baselworld 2016. One of them was the 36 mm model with 12 different stones, while the other was the 40 mm model with three different stones forming a series that was repeated four times with each revolution of the dial.

Stones 40 mm

Shinji Himeno

In November 2015, I received an email from a Japanese man, attached to which was a photo of a watch whose magnificent dial was engraved, in bas-relief, with a mythological scene. A fan of my work, he was hoping to collaborate with me. Ordinarily I refuse such proposals. I like to work on my own. However, sometime later, I received a parcel from Germany sent by a certain Shinji Himeno—the Japanese person in question. Inside was a compact disc with images of his work.

Shinji Himeno is a surrealist painter. His colours are those of the Renaissance. The perfect finish of his paintings and their extraordinary force of expression make them stunningly beautiful. I am very critical of contemporary art and abstract painting. On the other hand, I adore the surrealism of Dali and Magritte. I melt before the colours of Renaissance paintings. Shinji Himeno's paintings had everything it took to delight me. My breath taken away, I wrote to him to tell him of my interest in a collaborative endeavour. Finally—an important detail—I was touched by the connection between a Japanese person living in Germany and an Italian in Switzerland.

Two days later, he flew in from Germany to visit my workshop in Morges. As I got to know him better, I saw in him, as in myself, the sensitivity and suffering of people who live in their own world. We shared a number of common elements. Given his interest in mythology, for our project we chose Kronos, the god of Time. As for the movement, it would have to offer my Wandering Hours display. Above all, the dial would have to do without hands in order not to detract from the artwork. The hour would be displayed in a window wandering about the dial to indicate the minutes, and the seconds by the rotating hourglass in Kronos' right hand. He would be flitting away, along with time, pursued by the young woman attempting to stop him.

We gave a preview of this watch at the Basel Fair in 2016. Its success was ample reward for both of us. Selling a work is not the most important thing.

My Wandering Hours mechanism, created 20 years earlier, found a new home that allowed it to preserve its original design. I was content to "set in motion" the painting Shinji had created. What made this watch special was its appearance, but more than that, it told a common human story, the meeting of two strangers. There was no creative work in the technical sense.

Painting by Shinji Himeno

2015

Kronos watch

Personnelle Y
Case engraved by Mr. Olivier Roux

XVIII

Rage

Let us return to the notion of injustice for a moment. In Naples, we had to endure Spanish dominance from 1443 to 1707. When I was a child, the story of the revolt of Naples in 1647 made a huge impression on me. Its incarnation was a simple fisherman and fishmonger, Tomasso Aniello, who led the uprising against the exploitation of the middle class, supported by the Spanish. The people, armed with sticks and left to their own primitive devices, defeated the Spanish army. It was carnage. Hundreds were killed. Once this fishmonger had won, they wanted to make him the ruler of Naples. To everyone's surprise, he refused directly before the viceroy and his court, the cardinal and his prelates, and the people. He undressed and, naked as the day he was born, declared, "I didn't do all this for the sake of power, but of freedom. So that the people could live in peace and no longer be slaves." A few days later, the viceroy's men assassinated him.

Tommaso Aniello is known to history as "Masaniello", a combination of his first and last names. This episode really hit home. I see something of his character in me—the need to rebel, the desire for independence. My philosophy has always been one of revolt against falsehood, a lie, and exploitation. My father's influence is apparent in this too. Wanting only to be myself, I have never desired to be Swiss. We should not deny what we are. We should remain free to be true to our birth. To call myself Swiss would be to betray my origins, my parents, my past and myself.

Horology, the Watch Industry and Journalism

I have always been at war with institutions. The watch industry is one of them. The very word is the summation of every constraint, every concession and every obstacle. It is the reverse of freedom and independence. To be sure, it is the mark of an artist to cultivate his freedom. In my case, however, it went further than that. Watchmaking has never been the defining factor in my life. It has never interested me in itself—

especially not as it currently operates, with its hierarchical restrictions, its call to submit to its dogmas and traditions, its jealousies and financial imperatives based on the supposed wishes of the public. This is the reason that all my life I have taken a stand against the powers-that-be. I have shunned the paths mapped out by the great watch manufacturers. Only freedom and independence permit us to truly create in accordance with our nature.

Too many young people who entered the profession after me have failed to understand these things. Sometimes I feel like I am the last of the Mohicans. I believe certain firms understood it well enough to try to silence me, well enough to buy off the little Neapolitan troublemaker. Excluded from Swiss inner circles, he proved by his creations that you can and must avoid the well-beaten path and foster a spirit of innovation that is so lacking in the large firms.

Fortunately, a certain rivalry has nonetheless begun to make itself felt. It does not necessarily spring from the world of watchmaking, but from the most diverse crafts. I have also seen a number of young people coming into the profession with the courage of innovation.

Stop for a moment to consider the current monstrosities, so admirable in their complexity and yet so fragile. All it takes is for a speck of dust to get into the movement and the watchmaker will spend days or even weeks dismantling, repairing and re-assembling it. One may be amazed by such mechanical wonders, but that is not watchmaking. The craftsmen and his hands have no part in it.

Or else, consider the Geneva Seal, regarded as the epitome of quality control. This institution, praised by all the media, certifies the quality of the watch principally in terms of the finish of its components and gears and its array of bridges. Now these criteria in no way measure the reliability or precision of a watch. There is no point buying a watch for the impeccable finish of its parts if its actual functioning has not been perfected.

In the history of horology, it quickly became clear that a well-crafted movement needed the greatest possible number of bridges in order to regulate the working of the moving parts. Another great advance occurred when the improved machinery allowed for the elimination of chatons in watches. Nowadays, various prestigious firms are happy to fashion a single bridge for a number of moving parts and are boasting about reintroducing chatons.

Formerly, journalists played the role of critics. These days many of them have forgotten the notion. The success of the watch industry has given rise to an unheard number of media outlets devoted to it, an unbelievable number of journalists who never tire of writing about the large firms. Thirty years ago, there was no more than one or two horological publications in any country—and not every country had that. Now the media cannot get enough of the watch industry. It pays. Of course, these hordes of journalists, for the most part, know nothing at all about it. They write on commission or in partisan fashion, to say the least. It is no longer a question of information. It is misinformation.

Personnelle C
with 2017 sapphire case

XIX

A Touch of Philosophy

To be successful, in my view, means living a successful life, which must be grounded in philosophy—a philosophy of preserving one's values and sense of ethics and always being in search of one's own truth. If, in the pages of this book, I could give any advice to a young person who is starting out in this profession or in a creative life, I would tell him never to accept a statement of any sort without questioning it. Doubt and a critical sense are essential. I would tell him, "If you have a goal, a path, pursue it according to your own will, even if you go wrong at times. And if you come up against a wall, go around it or break it down, but don't allow yourself to be influenced by others. Otherwise, it would no longer be your path."

It had often been suggested to me that I should write a book about my life. It was, in fact, a book that guided my philosophical orientation. I was young, around 13, and still in Naples. My interest in philosophy increased with *The Power of the Will* by Paul-Clément Jagot, a French writer and occultist of the 40s. This work increased my strength of will tenfold, along with my intransigence and my total refusal to compromise or to give up. While it taught me how to master my emotions, it nonetheless prevented me from showing them. It was a long time before I understood this and began allowing them to express themselves.

Tourbillon C enamel

XX

An Eye on the Future

In the past, a watch was a timepiece, an object that told the time. Today, that function has become symbolic. For many people it evokes the quaint charm of one of the last artistic crafts; for others it has become a status symbol, even an investment. When we think about watches, we think about luxury—which is, of course, Swiss. In fact, this luxury industry is growing in both Japan and China. The Chinese are manufacturing ultra-complicated mechanical watches that are scarcely inferior to their Swiss counterparts. The Chinese and Japanese have stopped imitating; they are now creating.

This is to say that the Swiss watch is no longer listed in the register of perfection. It has lost its place as the symbol of quality. What is harder to admit is that it is no longer even truly Swiss. Fewer and fewer of the owners of large firms are Swiss. The same is true of their staff. Look at Swatch, Richemont and Vuitton. Even the turnover of the Swiss industry meets with derision—22 billion, compared to 27 billion for Migros and Coop. These two firms earn more than the entire Swiss watch industry, which in 2015 suffered a 16% downturn in exports. Their principal markets, China, Japan and the US, are buying less and less. What can one say? Since everything else on the planet is in trouble, there is no reason why our industry should be in good shape. Only the pharmaceutical industry is thriving: health still pays. This is because this industry meets a real need, which is not the case with watches and luxury items.

I cannot see exactly how to improve the situation. Everything depends on foreign demand, which depends on the global economy. There is much talk now about connected watches. I personally see no future in that. They tend to be too difficult to read, even with a very large watch. At best, they could serve as an alert device, but other such devices are available.

It helps to have a bit of perspective. The history of horology is shorter than we think. The oldest known clock was made in Padua in 1320. (There is a replica of it in the museum in La Chaux-de-Fonds and another in the Birmingham Museum.) In Renaissance times, much research took place in this field. There were developments in watchmaking in Germany (with the famous Nuremburg Egg), then in German-speaking Switzerland around 1500, before activity shifted to France and to England at almost the same time. As for Japanese watchmaking, it has been around since 1400 or 1500.

Without going so far back, we can note that owning a watch was always a privilege of the leisured classes. In the 19th century, with the advance of industrialization, and up until the 1950s, a watch was a prestige item. With quartz, it became much more commonplace. Furthermore, watchmaking was a family business. People were looking to the future, to leaving something for their children. The disaster of the 70s with its rollcall of bankruptcies was a boon to buyers, investors who were less interested in the future than the present. They had their own notion of time!

The early 80s saw the watch recover its sentimental and cultural value in grand style. It again became a cult item with its partisans, some of them fanatical, becoming a status symbol that made the market explode. This new infatuation turned the watch industry into a worldwide stock exchange. People bought watches, even factories, as if they were buying shares. The industry's turnover rose from some three billion to six, ten, even 20 billion. But the market had its moods, and gains alternated with losses. Hence all the reorganizations, mergers, managerial dancing about, the creation of the CEO and the advent of the marketing miracle.

The Basel Fair, or Baselworld

When I discovered the MUBA Trade Show in 1972, luxury had already made its appearance. Would-be exhibitors found participation affordable however, and tickets cost visitors less than ten francs. The big change occurred in 1990, when Cartier no longer wanted to share an exhibition with sausage vendors. The firm left Basel to create its own luxury showroom in Geneva—the *Salon International de la Haute Horlogerie (SIHH)*. Other companies followed, creating an opening for the return of the SMH group (later Swatch), which had earlier abandoned the fair.

Why this fierce rivalry between the Basel Fair and the SIHH and who was the real victim? Traditional watchmaking of course, more commonly called the Swiss watch industry! In the past, of the 2,000 exhibitors more than 600 were Swiss. Today it is only about 300 out of a total 1,500 exhibitors. In the past, you could meet all the watch manufacturers at the Basel Fair, as well as their suppliers, since the price of stalls was manageable. Bit by bit, we witnessed the exclusion of the smaller participants. Prices went up, tickets for the public rising from 10 to 60 francs.

In my opinion, Baselworld is the biggest enemy of the Swiss watch industry. Having become a financial powerhouse, it has forgotten its primary role as a global showcase for the watch industry. It no longer exists or acts except in the interests of the luxury-watch firms that can pay the vastly overpriced rents. In short, it has become associated with the MCH Global group, first listed on the Swiss stock exchange in 2001. The cost of living in Basel during the fair has sharply increased, some property owners increasing rents fivefold.

Thus Baselworld, like the Federation of the Swiss Watch Industry, has become an organization serving only providers of luxury items. The proof? The stalls on a royal scale. I pointed out to the organizers of the fair that the way they were going, they would put an end to all diversity. Why not devote one building to the luxury side of the industry, and another to its more modest side simultaneously? After all, without small enterprises there cannot be any big ones!

The Other Players

Limitless demand during good times led suppliers to increase their prices so much that they became prohibitive for small manufacturers. In addition, hard times, especially the crisis in 2008, caused many bankruptcies and takeovers of subcontractors by the large groups. Supply problems thus became even more serious, leading, by way of a chain reaction, to further bankruptcies among small firms. People would proudly talk about "Swiss Made"—but they did not hesitate to find suppliers in China, Korea or Taiwan. Skills and expertise have become decentralized. The irony is that those countries have outstripped Switzerland as regards the quality of their products, since qualified labour is scarcely to be found here anymore or else is simply too expensive.

When we consider collectors, the situation is hardly better. Some of them might own hundreds, if not thousands, of watches. Yet very few will have dealt directly with a true craftsman-watchmaker. That is the result of our marketing strategy. The Rolex Daytona is the most sought-after watch in the world simply because Paul Newman had one. It costs three or four times more than an ordinary Rolex. People have paid thousands of francs for the dial alone.

A large watch manufacturer is no longer capable of producing genuine, finely crafted watches. They no longer know how to set up a truly creative method of production. They are restricted by the realities of the industry: if their financial position requires them to shed employees, they lose some of their long-standing authenticity.

A brief word on counterfeit watches. The large firms laugh them to scorn—but there is no reason for this. After all, what is a counterfeit but reasonably priced replica of an overpriced original? The difference between a genuine item and a fake is that the fake might work as well as the original. I am exaggerating of course. Still, we have to admit that counterfeits are not necessarily of poor quality. And we have to ask the question: who are they stealing from? For at the end of the day, a counterfeit is advertising for the original.

Having said all this, let us not cast stones at the large firms. The problem is a global one. It is inherent in a system that tends to lose sight of its traditional values, and it is inevitable in an economic system that is based solely on the profit motive. It is not watchmaking itself that has gone astray, but rather the environment in which it functions. The consumer is bound above all by market forces. I recall the attitude of those unemployed men in Naples when I was a child, who would pay a king's ransom for a ticket to the upcoming soccer match while their families were slowly starving.

XXI

My Two Lives

My first life was before I became a creative watchmaker. My second was after. Before, I was a normal person like everyone else, firmly in pursuit of appearances and getting noticed. Then, one fine day, I felt the need to give something to others, to share a bit of my philosophy in a language they could understand. That language—my language, my means of expression—is watchmaking.

My occupation as a keeper of time has not changed my relationship with time itself. It remains the same as when I was 13, when I decided to live rather than die—and to live as fully and intensely as possible. I had not yet fashioned a system of ethics for myself. My rebellion was only latent. Yet I passionately embraced the unbroken thread of time. This intensity has not diminished with the passing of years. However, bit by bit it has transformed itself into a creative and philosophical quest.

To live for the sake of living has always struck me as absurd. That amounts to boredom and downright submission. When one becomes an actor instead of an onlooker, time loses its hold. When I was at play as a little boy, my mother would say, "That's enough! Go to bed now!" And when I was married, in the midst of my creative research, my wife would say, "That's enough! Come to bed now!"

The sort of time I love, introspective and liberating, cannot endure any outside interference. Is not time merely an abstraction, an ever-elusive convention? I have spent my life forging tools with which to measure it. Yet it remains elusive. I have accepted the word from on high. Unflinchingly. Like a good Neapolitan. This does not mean I am laying down arms, however. There is always a way to get around an obstacle, to fool it, to rise above it.

Some think that time is a universal, that it is our master. I have learned that it is relative, that it depends entirely on us. The struggle is worth it, even if it does not last long. A life can flash by in a second, and a second can last an eternity. Our master thus becomes our slave.

Were I To Do It All Over Again

I would do it the same way—because every time I undertake something, it is in a reasoned, analytical fashion. If I act in such-and-such way at such-and-such moment, it is because I know what led me to this moment of decision. Sometimes we think we should have done things differently. And yet it was a sequence of events that got us to where we were. My conclusion is that it is both stupid and useless to have regrets because they can spoil a life. Maybe I should have made certain concessions, not have cut my prices, or been more demanding, less tolerant, more of a salesman, less modest, more cunning. Now well beyond shame, I think I have reached the place I wanted to be.

I am not referring to my creations, but my life. My creations merely embody the language in which I can best express myself. They are simultaneously a need and a tool—a means of mobilizing my mind for the pursuit of my objectives, philosophical and material. It should be clear that I have chosen inner success over its financial counterpart. To be rather than to have. Money, even more than time, is a wicked master, because to own is to be owned. I have given my creations the generic name of Mechanical Poetry, because poetry is a means of expressing inconvenient truth.

To no longer push myself, to no longer be obliged to create in order to live—nor the reverse, for that matter: at the end of the day, is this not the highest challenge? Everyone, at some point or other, realizes that a bad grain of sand can block the flow of a life. I have experienced such a moment. One morning, while I was doing my usual exercises, clinging to a bar two metres off the ground, the bar gave way and I fell to the floor. Nothing was broken, but it was quite a shock. I thought it a miracle that I was still in one piece, but it was even painful to move. At the age of 70, this type of accident could have left me disabled or even been fatal. For me, it was a wake-up call. My family was right: it was time to write the story of my life. The life of a happy man.

Biography

1977

Buoyed by the gold medal awarded him at the International Exhibition of Inventions of Geneva, Vincent Calabrese undertakes his first creations, his Spatial watches.

He sells his patent to Corum, along with the manufacturing and exploitation rights on his linear movement utilized in his Spatial watches.

This step will lead to the invention of the "Golden Bridge", one of the most legendary watches in the history of horology.

1985

Vincent Calabrese founds the Académie Horlogère des Créateurs Indépendants (AHCI).
He creates his first wristwatch with a tourbillon, as well as the representative model of his Spatial collection, the "Esprit".

1986

Creation of the "Flying Tourbillon" for Blancpain.
This timepiece is still the thinnest watch with a tourbillon and the smallest eight-day movement ever produced.

1986-88

Further developments are made to the Spatial Collection, employing the most diverse metals for the creation of movements of various designs. Vincent Calabrese learns to craft on his own every part of a watch, from the movement to the case, including the enamelling, engraving and setting.

1988

Introduction of the "Flying Tourbillon", a supreme example of the principle of synthesis.

1989

Vincent Calabrese accepts the challenge issued by an Italian magazine to produce ten flying tourbillons within 18 months.

He begins to diversify, creating the "Two Hands" clock, which combines the immobility of a clock with the mobility of a tourbillon.

He also creates the "Baladin", which combines analogue and digital display with a jump-hour function.

While continuing to create successful watches for other manufacturers, he decides to produce a collection with movements of his own invention.

1990

Introduction of the "Commedia" jump-watch, of which in 1991 a new version named the "Mona Lisa" is introduced, thus becoming the first watch with a "jump-image" display.

1992

Creation of the "Power", very thin and highly reliable with power reserve indicator, which will lead to the development of his "Technique" collection.

A feature of this collection will be additional display functions, all of them regulated in various ways by the sole means of the winding-crown.

One model in this collection, the "Transworld", features a dual-time function, with local time being displayed in 12-hour format and home time in 24-hour format.

1993

Creation of the "Daily", a dual-time watch with a sub-dial indicating the weeks of the year.

1994

Creation of the "Mobil", a dual-time watch with a unique display.

1995

Creation of the "Night & Day", which automatically changes the hour display at noon and midnight, using Roman numerals for the hours before noon and Arabic numerals for the hours after.

1996

Vincent Calabrese reacquires the rights to his Baladin and embarks on the creation of his "Ludique" collection.

To celebrate his 52nd birthday, he creates the "52", a watch that offers of whole array of information necessary for day-to-day life, including the date, days of the week, weeks and months, which make this timepiece a delight to behold for collectors, connoisseurs and laypeople alike.

For his Spatial collection he creates the "Archipel du Japon", the "Equus", which displays a rearing horse, the "Regulus", a regulator tourbillon watch, and many other models.

1998

The "Horus" joins the Ludique collection, a watch with a novel Wandering Hours display, housed in a new case, also designed by the creator, with narrow lug-width.

Two new models are created to celebrate the millennium: firstly, the "Vincent", a phonetic pun on its creator's first name and on the figure "Vingt-Cent" (i.e. "twenty hundreds", or 2000).

It features a sapphire crystal that creates an optical effect, as tilting the watch causes the inner parts to disappear.

The case is in the shape of a flying saucer, resulting in a sleek and "playful" appearance.

And secondly, the "Two Times", with a novel mechanism allowing the hours and minutes to be displayed twice, as well as large date and week of the year functions.

This watch was produced as a limited series of 20.

2001

Creation of one of Vincent Calabrese's most daring watches, the "Sun-Tral", with jump-hour display in an aperture in the dial centre.

2002-2004

A whole array of devices produced for various watch manufacturers. A productive creative period giving rise to personalized timepieces in the spirit of his Spatial watches.

2004

Establishment of a partnership to create a personal trademark, Nouvelle Horlogerie Calabrese: At the Basel Fair he introduces a collection of four models, two of which are world premieres.

2006

Blancpain commissions a project involving overturning misconceptions concerning the differences between a tourbillon and a karussel. By creating a karussel that makes one revolution per minute and has a central balance wheel, Vincent Calabrese demonstrates that Bonniksen's invention was far superior to Breguet's.

2007

Vincent Calabrese sells his commercial interests to Blancpain and agrees to head a research centre in his Lausanne premises.

2008-11

Many innovations within the framework of the projects undertaken for Blancpain. Patent applications will be filed on nine of these.

2012

Vincent Calabrese again begins working independently and undertakes to promote the work of his friend, Jean Kazès, sponsoring his creation of sculptural clocks.

BIOGRAPHY

2014

Production of a special clock for L'Épée, the beginning of a new period of creativity.

2016

The Stones, a watch on which magnificent 'stones', chosen by the purchaser, tell the time.

This mechanism, the latest invention and creation by Vincent Calabrese, is unique: a window in the dial reveals a gorgeous stone!

Every hour, the dial moves round one twelfth (30 degrees) and displays another stone.

The client may choose either twelve identical stones or twelve of a different colour, material or weight.

Any combination is possible when creating a Stones, ensuring a truly personalised timepiece.

The watch movement can be manual, automatic or even quartz—in other words, made to the client's specifications—and its autonomy can vary from 36 hours to several years.

The Stones case can be 32, 34 or 40 mm and made of a material chosen by the purchaser.

Made to order only.

2016

Kronos, a magnificent symbol of time by two exceptional artists.

The work by Shinji Himeno, a painter of great talent, shows Kronos, the Titan god of time.

He is flying, clasping hours and minutes in one hand, seconds in the other.

A young woman is chasing time in vain, attracted by eternal youth.

Born in 1966 in Tokushima, Japan, Shinji Himeno studied at the *Hochschule für Grafik und Buchkunst* in Leipzip, Germany, and with Arik Brauer at the *Akademie der bildenden Künste* in Vienna, Austria. He has lived and worked in Berlin since 1997.

The mechanism of the Kronos brings the scene to life with one rotation per hour for the two characters and one per minute for the hourglass.
The position of the window indicates the minute, whereas the number indicates the hour. At the 60th minute, the hour showing gives way poetically to the next hour.

2017
New sapphire case

Glossary

Aperture (window)
Dial opening displaying various items of information—the hour, day, date, etc.)

Balance (balance wheel)
Vibrating organ that oscillates at a certain frequency, thereby dividing the flow of time into precise and equal segments.

Barrel
Wheel composed of a toothed disc and a cylindrical drum. The barrel contains the wound mainspring which, as it unwinds, supplies the power required to drive the watch.

Cabinet
Clock exterior, interpreted in many designs and materials, serving as both a protective and decorative element for the clock movement.

Chaton
Small, round gold or brass setting into which a watch movement jewel is inserted.
A historical technique rendered obsolete by current manufacturing methods.

Complication
Any additional mechanism designed for a function other than timekeeping, such as a dual-time indication, date display, etc.

Dial

Metal or other surface displaying various indications, generally the hours and minutes, although other indications are often also displayed, either on sub-dials or in apertures.

Differential

Mechanism linking two separate regulating systems that are turning at different rates.

Display

There are two sorts of displays—analogue, with the customary hours, minutes and seconds hands; and digital, with ordinary numerals.

Ebauche (blank)

Semi-assembled watch movement comprising various parts, such as the mainplate and bridges, onto which the other movement components are fitted.

Escapement

Mechanism designed to distribute energy from the geartrain to the regulating organ (balance wheel or pendulum).

Flying tourbillon

Tourbillon whose cage is pivot-mounted on one side.

Horns (lugs)

Variously shaped parts of a watch case to which a wristband may be fitted.

Isochronism

Property of the regulating organ of a watch, e.g., the balance wheel, pendulum or quartz crystal, whereby the period of its oscillations remains uniform.

GLOSSARY

Jump-hour display
Function whereby the hour is displayed by a numeral appearing through a dial aperture.
The change of hours occurs every 60 minutes, with the dial instantaneously or semi-instantaneously jumping 30 degrees.
The minutes are indicated by the usual sort of hand, either with the usual circular motion or as a retrograde display.

Karrusel
A mechanism similar to the tourbillon and created for the same purpose. Invented by the Danish watchmaker Bahne Bonniksen, it operates on the principle of the differential and has no fixed wheels.

Large date
Display through apertures fitted on two or three disks, depending on the complication.

Maillechort (nickel silver, German silver)
Alloy usually composed of 60% copper, 20% nickel and 20% zinc. Features a silvery appearance and is less liable to rust than brass.

Mainplate
Plate bearing the various movement components.

Marker
Reference point on a watch dial indicating the division of time in hours, minutes, weeks, months, etc.

Mirror (specular) polish
Perfect polish endowing a surface with the intense reflectivity of a mirror

Pendulum
The regulating organ in clocks.

Pivot
One end of the arbor (staff) of a moving part that spins in a hole in a plate that supports it.

Power reserve
Indication of how much stored energy remains in the watch and hence its remaining operating time.

Pusher (pushpiece)
Button on the watch case serving to adjust various additional functions.

Sub-dial
Small dial occupying part of the main dial.

Tourbillon
Mechanism that houses all the components of the escapement in a cage that generally spins at the rate of once a minute.
This rotation is achieved by means a fixed wheel and is based on the satellite gear principle.

Tourbillon cage (carriage)
A rotating part that houses the escapement and balance wheel.

Wandering Hours
Numerical display of the time through an aperture positioned around the rim on a mobile dial performing one rotation per hour. As with a jump-hour display, the hour changes instantaneously or semi-instantaneously at the 60th minute.
Minutes are indicated by the angular position of the aperture as it rotates around the dial.

Winding crown
Knurled or fluted knob used to rewind the watch or to set the time.

Acknowledgements

Words are powerless to express just how grateful I am to my daughters, to Evelyne, and to my family and many friends.

I do not name any of them to avoid oversight with the notable exception of Serge Bimpage, a man of letters who helped me with the writing of this book.

I would also like to thank in particular all those who have placed obstacles in my way and thus helped me to become the man I am today.

The website of Vincent Calabrese:

www.vincent-calabrese.com